Perseverance and the Mars 2020 M

Follow the Science to Jezero Crater

Manfred "Dutch" von Ehrenfried

Perseverance and the Mars 2020 Mission

Follow the Science to Jezero Crater

Manfred "Dutch" von Ehrenfried
Cedar Park, TX, USA

SPRINGER-PRAXIS BOOKS IN SPACE EXPLORATION

Springer Praxis Books
Space Exploration
ISBN 978-3-030-92117-0 ISBN 978-3-030-92118-7 (eBook)
https://doi.org/10.1007/978-3-030-92118-7

Project Editor: David M. Harland
Cover Design: Jim Wilkie

This Springer imprint is published by the registered company Springer Nature Switzerland AG
The registered company address is: Gewerbestrasse 11, 6330 Cham, Switzerland

Other Springer-Praxis Books by Manfred "Dutch" von Ehrenfried

Stratonauts: Pioneers Venturing into the Stratosphere, 2014
ISBN: 978-3-319-02900-9

The Birth of NASA: The Work of the Space Task Group, America's First True Space Pioneers, 2016
ISBN: 978-3-319-28426-2

Exploring the Martian Moons: A Human Mission to Deimos and Phobos, 2017
ISBN: 978-3-319-52699-7

Apollo Mission Control: The Making of a National Historic Landmark, 2018
ISBN: 978-3-319-76683-6

From Cave Man to Cave Martian, Living in Caves on the Earth, Moon and Mars, 2019
ISBN: 978-3-030-05407-6

The Artemis Lunar Program: Returning People to the Moon, 2020
ISBN: 978-3-030-38512-5

Stratospheric Balloons: Science and Commerce at the Edge of Space, 2021
ISBN: 978-3-030-68129-6

Evolution of a Martian

This artist's illustration shows NASA's four successful Mars rovers: Sojourner (on the left), Spirit & Opportunity, and Curiosity. Next, there is the Mars 2020 Perseverance rover and the Ingenuity Helicopter that arrived at Jezero Crater on February 18, 2021. Also depicted is a concept for how the NASA Mars Ascent Vehicle could be launched from the surface of the planet carrying tubes of rock and soil samples for a future Mars sample return mission. And finally there is a human explorer. Image courtesy of NASA

Dedication

It is hard to believe how many people have devoted themselves to the study of Mars; yet no one has ever been there! What is it that attracts so many to work decades, even their entire careers to gain a little bit more knowledge about an object over a hundred million miles away? When I look at Mars in the evening sky it is a mere dot on a black canvas. While many cannot imagine working on such a project, there are untold thousands that can't imagine pursuing any other type of work.

In conducting the research for the book, I was amazed at the caliber of scientists and engineers working on the Mars 2020 Program, now an operational mission. They come from all over, and from many different backgrounds and disciplines. They are so lucky to be part of the program and to work in such a unique place, namely the Jet Propulsion Laboratory (JPL) which is operated jointly by NASA and the California Institute of Technology (CalTech) in Pasadena. There are of course many others working on the Mars 2020 mission at other NASA Centers and at the establishments of our international partners. They too are dedicated to the program and the mission.

This book is therefore dedicated to the many different "teams" involved in the mission, including the science and engineering teams, the Perseverance team, the Principal Investigators, the Ingenuity team, the test and checkout team, the launch team, the cruise team, the EDL team, the surface team, the sample team…the list goes on. Having been part of an operations team in my distant past, I'm partial to those who're operating Perseverance and Ingenuity on the surface of Mars. I take it back, I'm envious!

So Go Mars Team Go, this book is dedicated to all of you. Enjoy it while you can, give it all you've got, and know that you are truly blessed to be part of something special, something grand, something historic! What stories you will be able to tell your grandchildren if the mission discovers life on another world!

Acknowledgments

Many thanks to those who reviewed my initial proposal to the publisher for their useful suggestions, and to those who provided input or comments on sections of the draft manuscript, in particular:

David C. Agle
Public Affairs Officer
NASA/JPL-CalTech, Pasadena, CA

Dr. Luther Beegle
PI of SHERLOC
Deputy Manager, Science Division
NASA/JPL-CalTech, Pasadena, CA

Dr. James F. Bell, III
Arizona State University
Tempe, AZ

Dr. Tanja Bosak
Professor of Geobiology
Massachusetts Institute of Technology
Cambridge, MA

Dr. Robert D. Braun
Director of Planetary Science
NASA/JPL-CalTech, Pasadena, CA

Dr. Svein-Erik Hamran
PI of RIMFAX
Professor of Radar Remote Sensing
Department of Technology Systems
University of Oslo, Norway

Dr. Michael H. Hecht
Associate Director for Research Management
MIT Haystack Observatory
Westford, MA

Dr. Christopher McKay
Astrobiologist
NASA Ames Research Center
Mountain View, CA

Brian K. Muirhead
Chief Architect and Pre-project Manager
Mars Sample Return
NASA/JPL-CalTech, Pasadena, CA

Dr. John McNamee
Mars 2020 Perseverance Project Manager
NASA/JPL-CalTech, Pasadena, CA

Dr. Roger C. Wiens PI
PI of SuperCam
Space & Planetary Exploration Team
ISR-2, Mail Stop C331, TA-00, SM1325
Los Alamos, NM

Charles Whetsel
Director, Engineering and Science Directorate
NASA/JPL-CalTech, Pasadena, CA

There are many other scientists whose reports I read in researching for this book and I have listed their works in the References section.

In addition, I acknowledge the assistance of Wikipedia and Google, and also the exceptional website of NASA/JPL-CalTech. These sites enabled me to fill in the pieces of the puzzle on just about any subject. Their inputs are woven into many sections.

Also, many thanks go to the people who have given me the opportunity to write about this unique program, particularly Hannah Kaufman, Associate Editor with Springer in New York, Clive Horwood of Praxis in Chichester, England, and the cover designer Jim Wilkie in Guildford, England. A special thanks to David M. Harland in Glasgow, Scotland who edited this, my eighth Springer-Praxis book. After over seven years of communications solely by email, I hope eventually to meet and thank him in person.

Thanks everyone, I hope you like the book and find it a handy quick reference.

Foreword

Philosophers and scientists have been struggling to understand our place in the universe for millennia. Finally, we are a step closer to obtaining real data that might just answer at least one of the questions. The Mars 2020's Perseverance rover is equipped with the tools to gather samples of the rocks and soils which will hopefully one day be returned to Earth so that we can directly investigate their biological potential. There are scientists who focus on what it takes for life to develop in the rocks and soils of distant worlds. These are the astrobiologists, geobiologists and others who probe the smallest of realms. One such scientist is Tanja Bosak, a professor of Geobiology in the Department of Earth, Atmospheric and Planetary Sciences at the Massachusetts Institute of Technology who is also involved in plans for analyzing samples of Mars that are returned to Earth.

The following is based on a statement Dr. Bosak made to the Subcommittee on Space and Aeronautics Committee on Science, Space and Technology of the US House of Representatives on April 29, 2021, titled: 'What do Scientists Hope to Learn with NASA's Mars Perseverance Rover?'

> How, where and when life originated is one of humanity's great unsolved questions. Until now, we were only able to explore this question on our own planet, where the record of life before 3.5 billion years ago is absent. In contrast, Mars has preserved a rock record up to a billion years older than the oldest well-preserved rocks on Earth. Studying these rocks will enable us, for the first time, to probe the emergence of life on another planet. This is the goal of the NASA Perseverance rover that landed in Jezero Crater on Mars. To achieve this, the mission aims to collect thirty samples of rocks and soils for future return to Earth as part of the international Mars Sample Return campaign.

Fig. 1 Dr. Tanja Bosak

Scientists will then be able to analyze these samples in laboratories. These will be the first set of samples from another planet that are relatable to specific locations and rocks or soils that were formed in established sequences of geologic events. Just as the Apollo program did for the Moon, these samples will revolutionize our understanding of Mars science and will stimulate enormous interest in science and technology for decades to come.

The scientific community has identified many questions that can be addressed by studies of the returned samples. These include looking for past life, tracking changes in the climate and atmosphere of Mars and its habitability through time, establishing when rivers and lakes existed, determining how impacts and weathering have affected the surface, and understanding the evolution of Mars' interior. Some of the returned samples will be collected in environments which were likely habitable during the time from which we have little record on Earth. For the first time ever, the scientific community will be able to look for the earliest stages of life by applying the criteria which were developed to investigate organic compounds and other potential signs of life in rocks from Earth.

> Analyses of returned samples can tell scientists what the earliest habitable environments looked like, whether early Mars received abundant building blocks for organic life from comets or asteroids, and whether some conditions at its surface enabled the synthesis of more complex organic compounds. Even more ambitiously, we can even ask whether any early life may have been transferred between Earth and Mars.
>
> The Perseverance rover has left its landing site to explore an ancient lakebed inside Jezero Crater. During this traverse, the science team has acquired remote imaging, radar and chemical data to characterize the geology of different regions at centimeter to kilometer scales and establish the time sequence in which the rocks were deposited.
>
> Instruments on the rover's arm have also imaged the rocks at scales comparable to those shown by a hand lens in order to determine the composition of minerals and look for potential signals of organic in rocks. Samples which are judged to have the greatest potential for answering the key science questions will be collected, documented and cached for return to Earth. The selection of a returnable sample cache is critical for the success of the subsequent legs of the Mars Sample Return campaign. It requires the identification of samples most likely to preserve organics under conditions that operated in habitable environments on early Mars.
>
> On Earth we can use a great diversity of instruments to determine the origin and ages of the Mars samples, reconstruct past climate change, characterize organic matter and search for signatures of past life. But owing to the small size of the collected samples, only small amounts of returned material will be available for analyses and most analyses will have to be performed at very small spatial scales, ranging from nanometers to millimeters. Nevertheless, such analyses can lead to a new understanding of the climate, the cycling of sediments, water, and inorganic and organic carbon on and within Mars. Any organic carbon present in the returned samples can also shed light on processes that control planetary habitability and lead to life. In the coming decades the returned samples will likely stimulate new developments at the intersection of geology, geochemistry, geobiology, materials science, mass spectrometry, microscopy, spectroscopy, planetary science, chemistry, and astronomy.

The following is taken from an article by Dr. Bosak in the March 2021 issue of Universe Today.

> The most direct test of the genetic relatedness of any Martian and terrestrial life would come from the comparisons of the information molecules (DNA and RNA) and the presence of such molecules in anything we find. In the best-case scenario we would find fossils of microbes or some such "biosig-

nature." Of course DNA and RNA do not preserve over billions of years (from the time when surface life was possible on Mars) but if we see something that looks like fossil cells upon sample return and detect some organic biosignatures, that would support there being similarities between past life on Mars and life on Earth.

Second, the discovery of evidence for past life on Mars is likely to lend some credibility to the theory that life still exists there today. Much like the disappearance of Mars' surface water, it is theorized that microbial life could have also migrated underground as a result of changes in the planet's climate. In fact, research has been conducted that demonstrated how microbes could survive beneath the surface in briny patches of water. The scientific consensus is that modern surface life on Mars is highly unlikely – which is why Perseverance aims to collect samples that will preserve evidence of past life. Nevertheless, the existence of past life will make the issue of planetary protection all the more pressing when human missions to Mars start, especially if we establish an enduring presence there.

Already, robotic missions are forced to exercise care in the vicinity of potential sites for microbial life, a good example of which is the time the Curiosity rover came upon a discolored patch of sand (thought to be a surface brine) and was forced to divert its path to go around it. If human habitats are ever built on Mars the possibility that we could be causing harm to Martian organisms will always be there.

The Perseverance rover will not provide the final word on this subject, but the data which it collects and the sample return it will perform will provide an essential piece to the puzzle. After all, the search for life on Mars is like the search for meaning in the universe: ongoing!

Dr. Tanja Bosak
Professor of Geobiology,
Massachusetts Institute of Technology,
Cambridge, MA

Preface

Since ancient times, Mars was at the forefront of astronomical observations as we struggled to understand our place in the universe. It was named for the god of war and was once thought to be a harbinger of death and plague. But then in the 19th century, telescopes trained on Mars by such observers as Giovanni Schiaparelli in Italy and Percival Lowell in the United States prompted imaginations to popularly envision an inhabited and vegetated planet.

However, with more advanced telescopic observations and the dawn of the space age, it was clear that Mars has a cold, dry environment that appeared to be devoid of life. The Mars 2020 mission is the first NASA mission to hunt for life since the Viking Program, almost a half century ago. As we continue to explore Mars, it is becoming increasingly apparent from orbiter and rover results that the planet was a very different place in the ancient past. Mars had periods of stable liquid water on its surface and for a significant period conditions were habitable. Orbital and in-situ observations show evidence of volcanic and hydrothermal environments that could have provided the energy required for life. The planet also had access to organic carbon, both from meteorites and by syntheses at hydrothermal vents. Dr. Luther Beegle, one of the Principal Investigators for Mars 2020, points out: "Three and a half billion years ago, Mars seems to have had everything the Earth did when life started on Earth, so, the question is did life start there? And if not, why not?"

"No evidence of life on Mars has ever been found. Each rover mission has inched closer to that goal, however," observes Dr. Abigail Allwood, an astrobiologist and Principal Investigator for one of the mission's instruments. Hence, the search for life continues with the understanding that science, and indeed humanity, requires an enormous burden of proof for the verification of ancient or indeed existing life on Mars. Consequently, it is not likely that NASA will declare that life has existed on Mars until the samples are brought back to Earth and analyzed in sophisticated laboratories.

The Mars 2020 robotic rover named Perseverance is even more advanced than its sister rover Curiosity, which has been roaming Gale Crater for almost a decade.

In some cases, Perseverance's instruments are a couple of orders of magnitude more sensitive than any instrument previously sent to Mars.

But this time it is different; much different. Earlier missions did not collect and cache samples; they were focused solely on investigating the surface to answer science questions about whether locations such as Gale Crater were habitable in the distant past. For Perseverance, on the other hand, the goal of collecting and caching samples changes the Mars 2020 team's approach to science exploration. This time the scientific exploration and analysis supports the innovative goal of caching those samples that best represent Mars as a planet and leaving them on the surface for a subsequent mission to collect and return to Earth for analysis.

One advantage of writing a book while the mission unfolds is that actual mission operations can be explained, presenting photos and videos of both successes and failures. But this book is also about the future, since it may be many years before the results of the Mars 2020 mission can be told. The samples that Perseverance leaves behind may not be brought back to Earth until the early 2030s. That joint NASA/European Space Agency program is known as the Mars Sample Return (MSR) mission and although it is described in this book, only the basic concepts and vehicles are presented. The current planning for this future program is for the Sample Retrieval Lander to be launched in 2028. The Earth Return Orbiter could launch in either 2027 or 2028 but the return of the samples, once planned for 2031 has already been delayed until 2033, and may well slip further. Only at that time will the detailed analysis of the samples in state-of-the-art laboratories be able to determine whether life existed, or still exits on Mars and what type of life is it?

As NASA Ames astrobiologist Dr. Christopher P. McKay states, "There may be life on Mars that shares a common ancestor with life on Earth. Hence the search for a 'second genesis' must focus on biochemistry and genetics. Only if we find evidence of life with a different biochemistry or genetics can we conclude that we've found evidence of a second genesis of life." What will that tell us about ourselves? Are we alone? If not, then what else is out there? Does life populate the solar system; the universe? What will it mean for humanity?

The current Mars 2020 effort began in 2013 with the development of the science community's wishes for Mars 2020 investigations and a series of workshops over a period of five years to decide on the best landing site to search for evidence of life, past and present. Needless to say, this was both a national and international quest for the optimum instruments and the optimum site involving thousands of scientists, engineers and mission planners. This book will describe that effort and the resulting definition of the Mars 2020 Goals, Science Objectives and Mission phases which guide the mission to this day.

In addition, much detail is included about the Perseverance design, components, stages and instruments. The seven major instruments are described in detail.

An Appendix gives the Principal Investigators for the instruments, along with other mission scientists and engineers. Also described are the flights of the helicopter technology demonstrator called Ingenuity.

On July 30, 2020 an Atlas V rocket launched NASA' Perseverance rover and Ingenuity helicopter toward Mars. They were tucked away in their aeroshell protected by the heat shield and guided by the cruise stage for over seven months. Like Curiosity nearly a decade earlier, Perseverance survived the famous "Seven Minutes of Terror" on February 18, 2021 and accurately and safely landed at the chosen landing site, on this occasion in Jezero Crater. Perseverance is presently exploring, probing and collecting samples in this ancient lakebed and the fan-shaped delta that fed it, seeking signs of past or existing life and collecting rock and soil samples. Some samples are to be sealed in super clean metal tubes and left on the Martian surface in the hope that a future mission will collect them and transport them to Earth for further analysis. The plan is for the mission to last at least a Mars year of 687 Earth days. If it is as successful as Curiosity, which has been operating on the planet for over nine of our years, the science community will be overjoyed.

While Perseverance has significant autonomous capabilities, such as driving itself across the Martian landscape, hundreds of Earthbound scientists are still involved in analyzing its results and planning further investigations. Although thousands of people have been involved with the mission over the years, there are almost 500 people on the current Mars 2020 science team. The number of participants in any given action by the rover is on the order of 100 people. It is an ongoing, real-time operation. Fortunately, JPL issues timely reports of the activities, many of which have been included in this book; for example, the approaches to selecting drilling sites and collecting samples are described. Also of interest is how the team have based their decisions on both Perseverance's instruments and data gleaned from the Ingenuity helicopter's views above and ahead of the rover's track. On April 19, 2021, Ingenuity achieved the first powered flight by any aircraft on another planet, an event as significant as the first flight at Kitty Hawk by the Wrights in 1903.

I hope you enjoy this description of the ongoing Mars 2020 mission. As it could go on for a decade or more, stay tuned for a possible sequel that further describes the successes and failures of NASA/JPL attempts to determine whether life once existed or currently exists on Mars and the ramifications of those discoveries for humanity.

Manfred "Dutch" von Ehrenfried
Cedar Park, TX, USA
Winter of 2021–2022

Contents

1

Introduction

A Brief History

Mars is similar to Earth in many ways, having many of the same "systems" that characterize our home world. Like Earth, Mars has an atmosphere, a hydrosphere, a cryosphere, and a lithosphere. In other words, Mars has systems of air, water, ice, and geology that all interact to produce the Martian environment. What we don't know yet is whether Mars ever developed or maintained a biosphere; an environment in which life could thrive. To discover the possibilities for past or present life on Mars, NASA's Mars Exploration Program is enacting a strategy called "Seek Signs of Life" that expands on the "Follow the Water" theme that motivated the Mars Global Surveyor and Mars Odyssey orbiters, the Spirit and Opportunity Mars Exploration Rovers, the Mars Reconnaissance Orbiter and the Mars Phoenix Lander (see Appendix 1).

Book Summary

This book chronicles the path along which the science community embarked to address its questions about Mars. It starts with the Viking missions launched in 1975 and works its way forward in time to describe what has transpired in orbit around and upon the surface of the planet, including the Perseverance rover and its Ingenuity helicopter.

Chapter 2 describes the goals and objectives of the Mars 2020 mission. This work by the science community for the past two decades culminated in a well-defined and concise set of statements which guide the Mars Exploration Program

M. von Ehrenfried, *Perseverance and the Mars 2020 Mission*,
Springer Praxis Books, https://doi.org/10.1007/978-3-030-92118-7_1

to this day. This chapter also defines the various mission phases which are necessary in order to finally arrive safely on the surface of Mars.

Chapter 3 provides the history of how the Curiosity rover design influenced the Perseverance design, to the point of copying some features and improving upon some others by applying the latest in technology. Because the rover design team had many years of actual experience with Curiosity operating at Gale Crater, the lessons learned were applied to the new Perseverance design vehicle. Similarly, the science community was able to learn from the performance of Curiosity's instruments and come up with a new set of seven instruments for Perseverance, some of which are far more sensitive and state-of-the-art and more relevant for this new mission. As the old saying goes, "This isn't your father's Oldsmobile."

Chapter 4 describes the many years of effort by the science community to agree on where Perseverance should operate. The Landing Site Selection Committee is described, as well as the many criteria and considerations that went into the final selection. After many workshops involving hundreds of scientists, the committee finally came up with the recommendation that the Mars 2020 Perseverance rover should investigate Jezero Crater, an area on Mars where the ancient environment may once have been favorable for microbial life. And sure enough, that is where Perseverance is to this day taking samples that will, some day, hopefully answer the key questions about Mars. The Mars 2020 entry, descent, and landing (EDL) phase employed the same "sky crane" landing system approach and design that successfully delivered Curiosity to the Martian surface. However, this time the system used several new technologies and features that enabled the spacecraft to land at previously inaccessible landing sites. A suite of cameras and a microphone captured both the sights and sounds of EDL for the first time, always an anxious and exciting seven minutes.

Chapter 5 describes the actual on-going operations, including the Perseverance rover moving to prime targets for sampling. The entire Sample Caching System and operating procedures are described, as are the actual sampling attempts and assessments. Where pertinent, commentaries by the science and operations teams commentary have been included, both failures and successes. Incredible pictures are included of the operations, rocks, samples and results. Activities were paused temporarily when the solar conjunction occurred between October 2 and 14, 2021. That was when Mars and the Earth were on opposite sides of the Sun. There is a moratorium on sending commands to Mars when the planet is within 2 degrees of the Sun from our perspective, as data could be lost.

Chapter 6 describes what was initially planned to be simply a 30-day technology demonstration of a helicopter, but turned out to be far more than expected. Once Ingenuity had established it could fly, the plan had been to abandon it so that the team could focus on the Perseverance mission. The plan was for Ingenuity to fly up to five times at heights up to 3–5 m (10–16 ft) above the ground, for up to

90 seconds each. On April 19, 2021 the helicopter achieved the first powered flight by any aircraft on another planet. After only a few flights, Ingenuity achieved its original objective of proving its ability to fly in the rarefied Martian atmosphere, over a hundred million miles from Earth without direct human control.

With Ingenuity's energy, telecommunications, and in-flight navigation systems performing beyond expectation, an opportunity arose to allow the helicopter to continue to explore its capabilities without significantly impacting on the rover's scheduling. It operated semi-autonomously, performing a variety of maneuvers planned, scripted and transmitted to it by JPL. Having established that powered controlled flight is possible on Mars, the Ingenuity experiment embarked upon a new phase of operations, expanding its flight envelope to include exploring how aerial scouting and other functions could benefit future exploration of Mars and other worlds. Along with those initial one-way flights, there was more precision maneuvering, increased use of its aerial-observation capabilities, and a greater overall acceptance of risk. This chapter chronicles all the flights during this first campaign.

Chapter 7 pays tribute to the various organizations at NASA Headquarters, JPL-CalTech and the international partners that contributed to the success of the Mars 2020 mission. Biographical outlines of many of the people in these organizations are given in Appendix 3.

Chapter 8 describes the follow-on Mars Retrieval Mission that is planned for the years 2026 to 2031. This mission will include a very complicated and technically challenging set of flights in order to retrieve the samples cached by Perseverance and return them to Earth. Also discussed is how samples will be disseminated to specially designed and equipped laboratories in which they will be analyzed and curated for future scientific research. Related to this are the Planetary Protection requirements and protocols.

In Chapter 9, I conclude with my assessment of the ongoing Perseverance and Ingenuity missions and my predictions for the future.

Appendix 1 gives an historical context for the Mars Exploration Program, going back to its origins in 1994 and explaining why the program was restructured in 2003. It discusses successes and failures along the way. Appendix 2 provides the conclusions from the Mars 2020 Science Definition Team. Appendix 3 provides many biographical outlines of the scientists and engineers of the mission, broken down into teams. Appendix 4 explains Dr. Christopher McKay's thoughts on the search for life on Mars, one of the major reasons for the exploration of Mars and indeed other solar system bodies. Other Appendices include notable quotes and a mission timeline up to the time of writing in late 2021.

It is envisioned that the Perseverance rover will still be roaming in Jezero Crater many years after this book is published. While you are hearing about events, you can use this book as a reference to better understand how they relate to the overall

Mars 2020 mission, and what will happen when the samples are returned to Earth. Like the Apollo lunar samples that are still being analyzed to this day by another generation of scientists, the samples collected by Perseverance over the next few years will be the focus of studies far into the future. Should life be discovered by this mission, who knows what the consequences will be for humankind!

2

The Mars 2020 Mission

The Mars 2020 mission with its Perseverance rover forms part of NASA's Mars Exploration Program, a long-term effort of robotic exploration of the Red Planet. The mission addresses high-priority science goals for Mars exploration, including key astrobiology questions about the potential for life on Mars.

The Perseverance rover is investigating a region where the ancient environment may have been favorable for microbial life. This involves probing the rocks for evidence of past life. Throughout its investigation, the rover will collect samples of soil and rock and cache them on the surface for potential return to Earth by a future mission.

Perseverance is carrying an entirely new subsystem to collect and prepare rocks and soil samples that includes a coring drill on its arm and a rack of sample tubes. About 30 of these sample tubes will be deposited at select locations for return by a potential future sample retrieval mission. Such specimens from Mars could be analyzed in laboratories on Earth for evidence of past life on Mars, and to assess possible health hazards for future human missions.

Two science instruments mounted on the robotic arm will be used to search for signs of past life. These will provide information to determine where to collect samples by analyzing the chemical, mineral, physical and organic characteristics of Martian rocks. On the vehicle's mast, two science instruments provide high-resolution imaging and three types of spectroscopy for remotely characterizing rocks and soil, in order to help in determining which rock targets to explore up close.

The Perseverance rover was provided with the same "sky crane" landing system developed for Curiosity, but with two enhancements to make more rugged sites eligible as safe landing candidates.

M. von Ehrenfried, *Perseverance and the Mars 2020 Mission*,
Springer Praxis Books, https://doi.org/10.1007/978-3-030-92118-7_2

2.1 SCIENCE GOALS

Background

NASA's Mars Exploration Program (MEP) had requested the Mars Exploration Program Analysis Group (MEPAG) to maintain the document named MEPAG Mars Science Goals, Objectives, Investigations and Priorities. This was initially released in 2001 as a statement of the Mars exploration community's consensus regarding its scientific priorities for investigations to be undertaken by (and in support of) the robotic Mars flight program. This document is regularly updated to respond to discoveries made by the missions of the Mars Exploration Program and changes in the strategic direction of NASA, the latest version being issued in 2020.

Historically, MEPAG has found that the pace of change in our knowledge of Mars is such that updates are needed roughly every two years. The MEP's intention is to use this information as one of its inputs into future planning, with no implied timeline for conducting the investigations. The rate at which investigations are pursued is at the discretion of NASA as well as other space agencies around the world that provide funding for flight missions. A separate, unrelated process for forward planning, similar in some ways to the Goals Document is the Planetary Science Decadal Survey which is prepared once every ten years by the National Academies of Sciences, Engineering and Medicine (NASEM), such as its Vision and Voyages for Planetary Science in the Decade 2013–2022.

The MEPAG Goals Document constitutes one of many inputs into the Decadal Survey discussion, even though these two organizations operate independently. The current version is again a four-tiered hierarchy of Goals, Objectives, Sub-Objectives and Investigations. The Goals are organized around four major areas of scientific knowledge, commonly referred to as: Life (Goal I), Climate (Goal II), Geology (Goal III) and Preparation for Human Exploration (Goal IV) all of which are described below. MEPAG does not prioritize among these four Goals because developing a comprehensive understanding of Mars as a system means making progress in all three science areas, and because the Goal of preparing for human exploration is very different in nature. Each Goal includes objectives that embody the knowledge, strategies and milestones needed in order to achieve the goal. The sub-objectives include more detail and clarity about different parts of objectives but cover tasks that are larger in scope than individual investigations. Hence, the investigations that go into collectively achieving each sub-objective constitute the final tier of the hierarchy. Although some investigations could be achieved with a single measurement, others require a suite of measurements, some of which require multiple missions.

Each set of investigations is independently prioritized within the parent subobjective. In some cases, the specific measurements that are needed to address an investigation are discussed; however, how those measurements should be made is not specified by the Goals Document, thereby allowing the competitive proposal process to identify the most effective means (i.e. instruments and/or missions) of making progress towards their realization. It should be noted that completion of all of the investigations in the MEPAG Goals Document would require decades. Given the complexity involved, it is also possible that they might never be truly complete, as observations which answer old questions often raise new questions. Thus, evaluations of prospective instruments and missions should be based upon how well investigations are addressed and how much progress might be achieved in the context of a specific instrument or mission.

Following the 2013 Announcement of Opportunities for the Mars 2020 mission further definition of the science goals and the design of the Perseverance rover, and even the instrument selection process, began with the establishment of the NASA Science Definition Team (SDT). The membership (selected by NASA from over 150 applicants) comprised scientists and engineers from a broad cross section of the planetary science communities whose areas of expertise included astrobiology, geophysics and geology as well as instrument development, science operations and mission design. The team was tasked to outline the Mars 2020 mission's objectives, realistic surface operations, a proof-of-concept instrument suite, and suggestions for threshold science measurements that would meet the proposed objectives. They were also to consider the Planetary Decadal Survey science recommendation for the highest priority "large mission" for the decade 2013–2022. The SDT effort occurred shortly after the successful landing of the Mars Science Laboratory's Curiosity rover on August 6, 2012 as the latest in a run of technological and scientific triumphs of the Mars Exploration Program. The SDT published its report on July 1, 2013.

Scientists and engineers on competitive teams focused on designing instruments to match the established criteria and goals. NASA selected optimal components from among the submitted proposals. It then openly competed opportunities for the specific payloads that would address the goals of the mission. The choice of the Mars 2020 rover science instruments was announced on July 31, 2014. The science instruments would support studies related to habitability, the search for any potential signs of past microbial life, identifying the most compelling samples for future potential Earth return and activities preparing for possible future human exploration.

In the following description of the science goals and objectives, the term "we" is used to collectively denote the MEPAG scientists who worked over the course of eight years to develop the Mars 2020 mission.

2.1.1 Goal I: Determine Whether Mars Ever Supported Life, or Still Does

The mission of the Mars 2020 Perseverance rover is focused on surface-based studies of the Martian environment, seeking preserved signs of biosignatures in rock samples that formed in ancient Martian environments with conditions that might have been favorable to microbial life.

It is the first rover mission designed to seek signs of past microbial life. Earlier rovers first focused on establishing that Mars once hosted habitable conditions. During the next two decades, NASA will undertake several missions to address whether life ever actually arose on Mars.

Conditions Needed for Life to Thrive

On Earth, all forms of life need water to survive. It is likely, though not certain, that if life ever evolved on Mars this occurred in the presence of a long-standing supply of water. On Mars, we will therefore search for evidence of life in areas where liquid water was once stable as well as below ground where it still might exist. There might also be "hot spots" on the planet, where hydrothermal pools (like those at Yellowstone) offer places for life. Recent data from Mars Global Surveyor hint that liquid water may exist just under the surface in rare places on the planet and the 2001 Mars Odyssey mapped subsurface water reservoirs on a global scale. We know that water ice is present in the Martian polar regions, so these areas will also be good places to search for evidence of life.

In addition to liquid water, life also needs energy. Future missions will therefore also be on the lookout for energy sources other than sunlight, because life on the surface of Mars is unlikely given the presence of "superoxides" that break down organic (carbon-based) molecules on which life is based. Here on Earth, we find life in many places where sunlight never penetrates: in dark ocean depths, inside rocks, and deep below the surface. Some forms of life on Earth exploit chemical and geothermal energy. Perhaps subsurface microbes on Mars could make use of such energy sources too.

Looking for Life Signs

NASA will also look for life on Mars by searching for certain "biosignatures" of current and past life. The element carbon, for instance, is a fundamental building block of life. Knowing where carbon is present, and in what form, would tell us a lot about where life might have developed.

We know that most of the current Martian atmosphere consists of carbon dioxide. If carbonate minerals were formed on the surface by chemical reactions between water and the atmosphere, the presence of these minerals would be an indication that water had been present for a long time, perhaps long enough for life to have developed.

On Earth, fossils in sedimentary rock leave a record of past life. Based on studies of the terrestrial fossil record, we know that only certain environments and types of deposits provide good places for preservation of fossils. On Mars searches are already underway to locate lakes or streams which may have left behind similar deposits.

So far, however, the kinds of biosignatures we know how to identify are those on Earth. Life on another planet may be very different. The challenge is to be able to differentiate life from nonlife no matter where one finds it, and irrespective of its chemistry, structure, and other characteristics. Life detection technologies under development will help us to define life in non-Earth-centric terms, so that we are able to detect it in all the forms it might take.

2.1.2 Goal II: Understand the Processes and History of Climate on Mars

Past Martian climate conditions are a focus of the Perseverance rover mission. Its instruments are looking for evidence of ancient habitable environments in which microbial life could have existed in the past. A top priority in our exploration of Mars is understanding its present climate, what its climate was like in the distant past, and the causes of climate change over time.

The current climate is regulated by the seasonal cycles of the carbon dioxide ice caps, the movement of large amounts of dust by the atmosphere and the exchange of water vapor between the surface and the atmosphere. One of the most dynamic weather patterns on Mars is the formation of dust storms that generally develop in the southern spring and summer. These storms can grow to encompass the whole planet. Understanding how these storms develop and grow is one goal of future climatic studies. An improved understanding of Mars' current climate will help scientists to model its past climatic behavior more effectively. To do this, NASA will need detailed weather maps of the planet and information about how much dust and water vapor are in the atmosphere. Monitoring the planet over one full Martian year (687 Earth days) will help us to understand how Mars behaves over its seasonal cycle and guide us toward understanding how the planet changes on timescales of millions of years.

The layered terrain of the polar regions also provides clues to the planet's past, in much the same way the rings of a tree document its history. When and how were these polar layers deposited? Was the climate of Mars ever like that of Earth? And if so, what happened to change the planet into the dry, cold, barren desert that it is today? Those are the questions that our missions still have to answer.

The Perseverance rover has a weather station known as the Mars Environmental Dynamics Analyzer (MEDA). It will help to study the shape and size of the dust in the atmosphere. This knowledge will shed light on how Mars' dust may affect human health. The instrument's measurements include wind speed and direction, temperature and humidity, as well as the amount and size of dust particles in the atmosphere. Such measurements will add to an understanding of the atmosphere as a whole.

To further our understanding of the temperature conditions in Mars' atmosphere, MEDA will also collect information during the entry, descent, and landing phase using a set of sensors attached to the heat shield and back shell to collect data as the spacecraft plunges through the atmosphere.

2.1.3 Goal III: Understand the Origin and Evolution of Mars as a Geological System

The Perseverance rover is designed to study the rock record to reveal more about the geological processes that created and modified the Martian crust and surface over time. Each layer of rock documents the environment in which it was formed. The rover seeks evidence of rocks that formed in water and that preserve evidence of organics, the chemical building blocks of life.

How did Mars become the planet we see today? What accounts for the differences and similarities between Earth and Mars? These questions will be investigated by studying Mars' geology. As part of the Mars Exploration Program, scientists want to find out how the relative roles of wind, water, volcanism, tectonics, cratering and various other processes have acted to form and modify the surface.

For example, Mars has some incredibly large volcanoes, 10 to 100 times larger than those on Earth. One reason for this difference is the absence on Mars of the process of plate tectonics by which the crust drifts over "hot spots." On Mars the immobile crust means the total volume of lava simply piles up to create massive volcanoes.

The discovery by Mars Global Surveyor of large areas of magnetic materials on Mars indicates that the planet once had a magnetic field, much like Earth does today. Because magnetic fields in general act to shield planets from many forms of cosmic radiation, this discovery has important implications for the prospects for evidence of past life on the Martian surface. The ancient magnetic field also helps us to understand the interior structure, temperature and composition of the planet in the past, suggesting Mars was once more of a dynamic Earth-like body than it is today.

Of fundamental importance are the ages and compositions of different types of rocks on the Martian surface. Geologists use the age of rocks to determine the sequence of events in a planet's history. Composition information tells us what happened over time. Of particular importance is the identification of rocks and minerals that were formed in the presence of water. Water is one of the keys to whether life might have started on Mars.

What other materials might be trapped in the rocks to tell us about the planet's history? How are the different rock types distributed across the surface? Future orbiting and landed missions will carry special tools designed to help to answer precisely these questions. Perseverance will attempt to answer at least some of them.

2.1.4 Goal IV: Prepare for Human Exploration

This goal also relates to national space policy for sending humans to Mars in the 2030s. Similar to the history of the exploration of Earth's moon, robotic missions to Mars provide a crucial understanding of the environment and test a number of innovative technologies for future human exploration.

Getting astronauts down to the Martian surface and then returning them safely to Earth is an extremely difficult engineering challenge. A thorough understanding of the Martian environment is critical to the safe operation of equipment and to human health, and the Mars Exploration Program will increasing address these challenges. Perseverance will hopefully contribute to that understanding.

The safety of astronauts will be of paramount importance because Mars is not a healthy place for humans. Mars lacks an ozone layer, which on Earth shields us from lethal doses of solar ultraviolet radiation. We do not have good information about the amount of ultraviolet radiation that reaches the Martian surface. A more detailed understanding of the surface environment will provide the data needed to design space suits and habitats to protect against this radiation.

We know the Martian soil hosts superoxides, and in the presence of ultraviolet radiation these break down organic molecules. While the effects of superoxides on suited astronauts will probably not be serious, their presence (and possibly a variety of unique chemical components of the soil) must be assessed before we can safely initiate human exploration of the planet.

To pave the way for human exploration, the 2001 Mars Odyssey orbiter has been providing data for twenty years. It and the Mars Reconnaissance Orbiter continue to search for water resources that, if discovered, could be used to support human explorers. Eventually, robotic spacecraft, rovers, and drills could be used to gain access to water resources in advance of (and indeed during) human exploration. Advanced entry, descent and landing techniques that will reduce the G-forces on landers have also been developed for spacecraft and astronaut safety.

While robotic exploration primarily related to Perseverance is headed by the Jet Propulsion Laboratory in Pasadena, California, much of the necessary scientific and technological work for the human exploration goal is carried out by NASA's Johnson Space Center in Houston, Texas.

Perseverance is demonstrating key technologies for using natural resources in the Martian environment for life support and fuel. It is also monitoring environmental conditions to enable mission planners to better understand how to protect human explorers.

Fig. 2.1 Humans on Mars. This artist's concept depicts astronauts and human habitats on Mars. The Mars 2020 Perseverance rover is carrying a number of technologies that could assist preparations for human exploration of the planet. Photo courtesy of NASA

2.2 SCIENCE OBJECTIVES

Missions preceding Perseverance established that liquid water existed on Mars in the ancient past and they explored the planet's "habitability." For example, at its landing site in Gale crater the Curiosity rover found the chemicals necessary for life (namely sulfur, nitrogen, oxygen, phosphorus, and carbon) as well as energy sources that microbes could have used, confirming there were regions that could have been friendly to life in the ancient past.

Perseverance will take the next natural step in Mars exploration. It will look for locations in Jezero Crater that were habitable in the distant past, ask whether life ever existed there, and look for signs of ancient life. The mission will also study the evolution of Mars' climate, surface, and the interior of the planet. The rover will evaluate technologies for future human exploration of the planet. In addition to these science goals, Perseverance has a unique goal of collecting and caching samples of Martian material for possible future return to Earth.

Because earlier missions did not collect and cache samples, those were focused solely on studying the surface to answer science questions. For Perseverance on the other hand, the goal of collecting and caching samples changes the approach to science exploration. The exploration and analysis that the team is undertaking directly supports the goal of caching those samples that best represent Mars as a planet.

Perseverance has four science objectives, as follows:

2.2.1 Objective A: Geology

Study the rocks and landscape at the landing site in order to reveal the region's history and characterize the processes that formed and modified the geological record in an area selected for evidence of an astrobiologically relevant ancient environment and geological diversity.

One of the goals of the Mars 2020 mission is to explore an area that once had the potential to host and preserve ancient life. Perseverance will explore a region that could answer key questions regarding the potential for ancient life on Mars: Was it warm? Was it wet? Was it hospitable to life?

This is why Jezero Crater was selected as Perseverance's landing site. The data from orbiters implies the crater was once home to an ancient delta flooded with water. In particular, there are clay minerals and carbonates which can form only in the presence of water. Carbonates are also a great material for "recording" an ancient climate. Terrestrial carbonates are commonly produced by life, and are known to preserve evidence of it. This made Jezero Crater a promising place to search for signs of ancient life.

To meet its first objective, Perseverance will:

- Study the area to understand its historical climate and environment.
- Study its geological diversity to determine the different types of rocks and minerals present.

This is important because carbonates and certain types of clays can be excellent minerals in which to look for signs of past life, whereas volcanic rocks are good candidates for determining the age of the location the rocks were obtained from.

This data will help piece together a larger picture of what Jezero Crater was like in the ancient past; in particular what the climate was like and whether the delta and lake settings were friendly to life and if so, for how long.

2.2.2 Objective B: Astrobiology

Determine whether an area of interest was suitable for life, and look for signs of ancient life itself. Perform the following astrobiologically relevant investigations on the geological materials at Jezero Crater:

- Determine the habitability of an ancient environment.
- For ancient environments interpreted to have been habitable, search for materials with high potential for the preservation of biosignatures.
- Guided by the indications of habitability and preservation, seek potential evidence of past life.

Once Perseverance has a chance to study the landing site and its history, the team should be able to identify whether any locations in the crater were once habitable with both liquid water and the chemical building blocks of life being present?

As the rover explores Jezero Crater, it will look for areas that were particularly friendly to life, and in such places the team will look for deposits that could have preserved signs of ancient microbial life, particularly clays and carbonates. Upon locating such materials, the rover's instruments will search for potential evidence of past life. These signs could include characteristic shapes in rocks, or "patterns" in rock chemistry.

2.2.3 Objective C: Sample Caching

Obtain the most promising samples that represent the geological diversity of the landing site and thoroughly document their context. Guarantee compliance with future needs in terms of planetary protection and engineering, so that the cached samples can be returned to Earth in the future if NASA should choose to do this.

Scientists working on Earth normally collect samples in person. While collecting a sample, it is essential to document the surrounding "context" of any samples, or the environment and surroundings in which it occurs. This way, when the samples are brought to the laboratory, anyone working with them can find out more about where they were collected and why.

The samples which Perseverance caches for later collection will become the first-ever Mars samples to be brought back to Earth. Since Perseverance doesn't have a human onboard, the JPL team must document the context in which each sample is collected, including information such as where the sample was taken, which types of rock were found above or below it, why it was collected, and anything else that might assist laboratory analysis.

If and when these samples are brought back they will represent the planet Mars, and hence should represent as complete a picture of the landing sites as possible. Different types of samples will help shed light on different aspects of the history of the crater. This is why the mission team must ensure that they obtain a variety of rock and soil types from within Jezero Crater.

As described above, we know based on observations from orbit that carbonates and clays are present in Jezero Crater. The clays and sediment brought in by the river were possibly deposited and buried in the ancient river delta. These could preserve biosignatures. On the other hand, the floor of the crater might contain volcanic rocks that would reveal the age of a location. Still other samples could shed light on the ancient climate of the planet. This is why the science team will set out to collect a diverse suite of samples at Jezero Crater.

The plan is to deposit samples at one or more suitable locations on the surface of Mars. The tubes with the samples were designed so that they can be brought back to Earth safely someday. If and when that happens, teams on Earth would have all the background information they require to understand where, how, and when the sample was acquired. Connections between the types of samples collected might reveal more about the environment in which they were formed, greatly adding to our understanding of the history of Mars as a planet.

2.2.4 Objective D: Prepare for Humans

Contribute to the preparation for future human exploration of Mars by making significant progress with at least one of the Strategic Knowledge Gaps that will have to be closed in order to sustain a human presence on Mars.

The highest priority measurements that are synergistic with Mars 2020 science objectives are:

- Demonstration of in-situ resource utilization technologies to facilitate the production of propellant and consumable oxygen from the atmosphere of Mars for future exploration missions.
- Characterization of atmospheric dust size and morphology to understand its effects on the operation of surface systems and human health.
- Surface weather measurements to validate global atmospheric models.
- A set of engineering sensors embedded in the Mars 2020 heat shield and back shell to gather data on the aerothermal conditions, thermal protection system, and aerodynamic performance characteristics of the entry vehicle during its descent to the Martian surface.

Note that many of the goals and objectives previously mentioned also relate to the preparation for future human presence, such as:

- The Mars Oxygen In-Situ Resource Utilization Experiment (MOXIE) on Perseverance will demonstrate a way that future explorers might produce oxygen from the Martian atmosphere for propellant and for breathing.
- The Mars Environmental Dynamics Analyzer (MEDA) weather station on the rover will help study the shape and size of the dust in the atmosphere. This knowledge will help shed light on how Mars' dust may affect human health. The instrument's measurements include wind speed and direction, temperature and humidity, as well as the amount and size of dust particles in the atmosphere. Measurements of the weather collected on the surface will contribute to our knowledge of the Martian atmosphere as a whole.
- To add to our understanding of the temperature conditions in the Martian atmosphere, the mission will also use MEDA sensors attached to the heat shield and back shell to collect data during its entry, descent, and landing phase.

- The Scanning Habitable Environments with Raman & Luminescence for Organics & Chemicals (SHERLOC) instrument also carries small pieces of spacesuit material to determine how they hold up in the harsh Martian environment.

2.3 MISSION PHASES

2.3.1 Pre-launch Activities

The pre-launch period covers everything from initial mission design to all stages of building and testing the rover, its spacecraft, and the launch vehicle. While the development engineers work on building and testing, scientists at universities and research institutions throughout the world plan their instrument observations and decide how to make the best use of the rover's capabilities.

Pre-launch activities include:

- Pre-Project Planning: Mars 2020 completes its concept and requirements definition (known as Phase A) and its preliminary design and technology development (Phase B).
- Science Definition and Instrument Selection: NASA selects a science team to propose mission objectives and desired capabilities, and then scientists propose instruments capable of achieving these objectives.
- Final Design and Fabrication at NASA's Jet Propulsion Laboratory: The Mars 2020 mission team builds the parts of the rover and its spacecraft (Phase C).
- Landing Site Selection: The Mars 2020 team considers more than sixty potential landing sites. The final selection is Jezero Crater.
- Engineers assemble and test the Mars 2020 rover in a clean room at JPL (Phase D).
- Shipping the spacecraft to NASA's Kennedy Space Center in Florida for assembly and testing.

2.3.2 Launch

A launch window occurs when the positions of Earth and Mars are optimal for a flight to Mars. In the case of Perseverance, the window opened on July 17, 2020 and lasted through to August 15. It was launched in the middle of the window, on July 30, 2020, at 7:50 a.m. EDT on an Atlas V-541 rocket from Launch Complex 41 at Cape Canaveral Air Force Station, Florida. The code "541" signifies a 5 m (17 ft) short payload fairing, four solid-rocket boosters and a one-engine Centaur upper stage. Developed by United Launch Alliance of Centennial, Colorado, and standing 60 m (197 ft) tall, the Atlas V is one of the largest rockets available for interplanetary flight.

Fig. 2.2 Mars 2020 launch. Photo courtesy of NASA

2.3.3 Cruise

The cruise phase started after the spacecraft separated from the rocket, soon after launch. The spacecraft departed at a speed of about 39,600 kph (24,600 mph) on a trajectory to Mars of 480 million km (300 million mi) taking about seven months. During that journey, engineers had several opportunities to adjust the spacecraft's flight path to optimize its speed and direction for arrival at Jezero Crater. The JPL team performed the following activities:

- Checking spacecraft health and maintenance.
- Monitoring and calibrating the spacecraft and its onboard subsystems and instruments.
- Performing attitude correction burns (to keep the antenna pointed to Earth for communications and to keep the solar panels facing the Sun to provide electrical power).
- Conducting navigation activities such as trajectory correction maneuvers to determine and correct the flight path and also to train navigators before atmospheric entry. The final three correction maneuvers were scheduled during the approach phase.
- Prepared for entry, descent, and landing (EDL) and surface operations, a process focusing on communications tests, including communications to be used during EDL.

Fine-tuning the flight path to Mars

The JPL team adjusted the path to Mars six times during the cruise phase. Each trajectory correction maneuver involved engineers calculating the spacecraft's location in space and commanding eight thrusters on the cruise stage to fire for the duration needed to adjust the path.

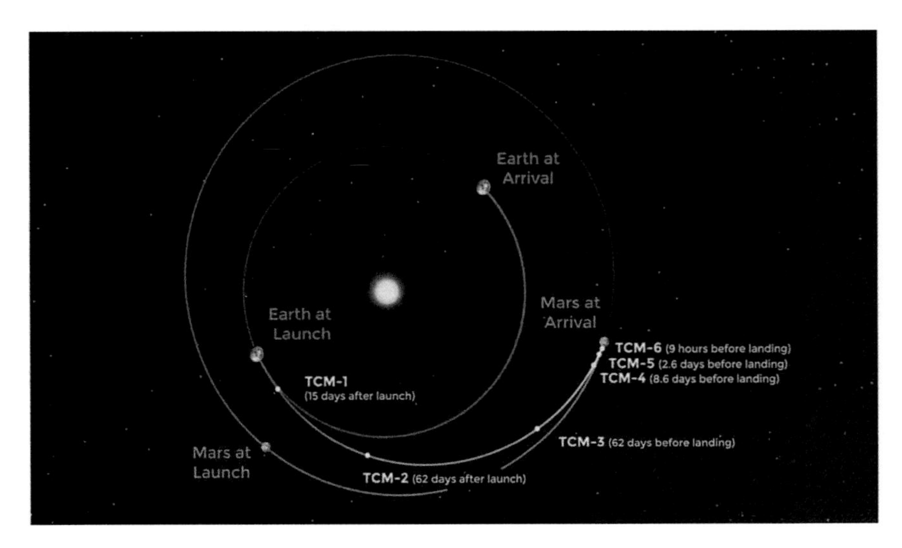

Fig. 2.3 Adjusting the trajectory to Mars. Photo courtesy of NASA/JPL-CalTech

2.3.4 Approach

To ensure a successful entry, descent, and landing, JPL engineers began intensive preparations during the approach phase, 45 days before the spacecraft entered the Martian atmosphere at an altitude of 3,522.2 km (2,113 mi) as measured from the center of the planet.

The team's activities during the approach phase included:

- The trajectory correction maneuvers to make the final adjustments to the spacecraft's incoming trajectory at Mars.
- Attitude pointing updates, as necessary to maintain communications and solar power.
- Frequent Delta DOR (Differential One-way Range; an interplanetary radio-tracking and navigation technique) measurements that monitor the spacecraft's position and ensure accurate delivery.
- Initiating the entry, descent, and landing software, which automatically executes commands during that phase.
- Entry, descent, and landing parameter updates.

- Spacecraft activities leading up to the final turn to the entry attitude and separation from the cruise stage.
- Uploading the surface sequences and communication windows for use during the first few sols.

Fig. 2.4 The approach phase. Photo courtesy of NASA/JPL-CalTech

During the approach phase, the mission team substantially increased the amount of tracking required by the Deep Space Network to allow engineers to determine more accurate trajectory solutions in the final weeks before arrival at Mars. This tracking supported the safe delivery of all the robotic spacecraft to the surface of Mars. The 34 m (112 ft) and 70 m (230 ft) antennas provided tracking during the approach phase.

2.3.5 Entry, Descent and Landing

The intense entry, descent, and landing (EDL) phase began when the Mars 2020 spacecraft reached the top of the Martian atmosphere traveling at about 19,500 kph (12,100 mph). About 10 minutes earlier the spacecraft shed the cruise stage. The spacecraft manipulated its descent into Mars' atmosphere using a technique known as "guided entry" to reduce the size of the targeted ellipse-shaped landing area while compensating for variations in the density of the Martian atmosphere and drag on the vehicle. In a guided entry, small thrusters mounted on the rear of the aeroshell modify the direction of aerodynamic lift to allow the spacecraft to control how far downrange it travels.

The peak of heating occurred about 75 seconds after atmospheric entry, with the temperature at the external surface of the heat shield increasing to about 1,300°C (2,370°F). About three minutes later Perseverance's parachute deployed with the help of a new technique named Range Trigger, which autonomously updated the deployment time for the parachute based on navigation position. It calculated the spacecraft's distance to the landing target and opened the parachute at the "ideal" time for the spacecraft to hit its mark. This resulted in a smaller and more precise landing ellipse, or target landing zone. The landing ellipse for Perseverance was 10 times smaller in area than that for the Curiosity rover in 2012 and almost 300 times smaller than that of the first Mars rover, Sojourner, delivered by the Mars Pathfinder mission in 1997.

At 21.5 m (70.5 ft) in diameter, the parachute was deployed about 240 seconds after entry at an altitude of about 11 km (7 mi), having slowed to a velocity of about 1,512 kph (940 mph). Twenty seconds later the heat shield separated and dropped away. This revealed a radar and cameras which fed into the other new landing technology called Terrain-Relative Navigation, a kind of autopilot that can quickly figure out the vehicle's location over the surface and select the best reachable safe landing target.

Previous Mars missions had relied on radar to help determine how far they were from the ground and how fast they were going during landing. Perseverance had data from a radar plus another new element known as the Lander Vision System. The radar actively "pinged" the ground from the time the heat shield came off at an altitude of about 7 to 8 km (4 to 5 mi) all the way to touchdown. The Lander Vision System (part of the Terrain-Relative Navigation technology) functioned when the spacecraft was between about 4.2 to 2.2 km (2.6 to 1.4 mi) in altitude. Terrain-Relative Navigation is able to adjust the point of landing by up to about 600 m (2,000 ft).

The Lander Vision System's job was to determine the rover's position above the ground in less than 10 seconds, handling different possible terrain conditions, to an accuracy of about 40 m (130 ft). It had a downward-facing camera which took image after image of the approaching ground and an onboard computer known as the Vision Compute Element which processed the images and spat out locations. After the camera activated, the Lander Vision System used an initial 5 seconds to take three images and process them to calculate a rough position relative to the Martian surface. Then, using that initial location solution it took further images and processed them every second to derive locations on a finer scale. The Vision Compute Element sent a stream of these location calculations to the main rover brain, called the Rover Compute Element. At about the same time that the rover separated from the back shell (with its parachute), the Rover Compute Element used the last accurate location calculation from the Lander Vision System to pick the safest reachable landing site, thereby ending the Terrain-Relative Navigation activities.

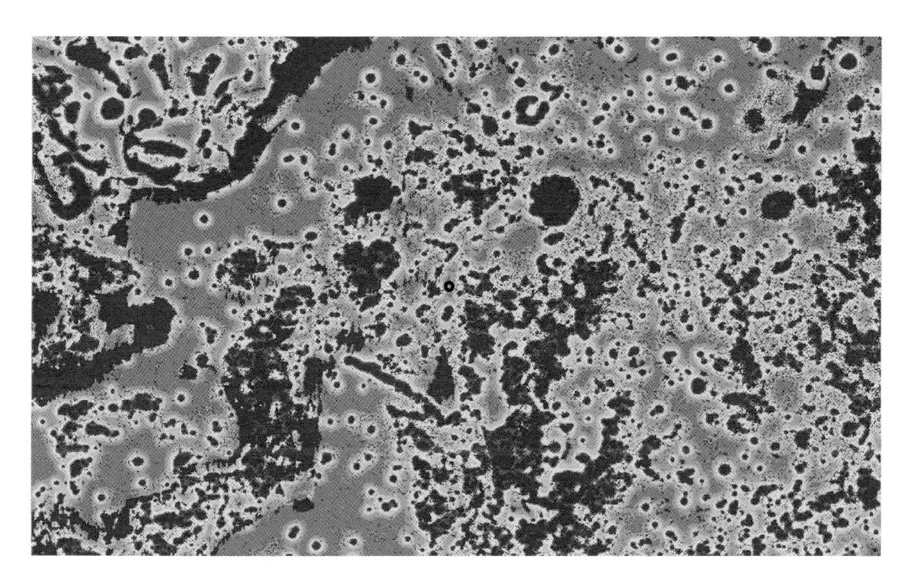

Fig. 2.5 Avoiding hazards at Jezero Crater. The areas shown in blue were considered safe zones and red ones more dangerous. The green dot indicates Perseverance's actual landing point. Photo courtesy of NASA/JPL-CalTech

When the descent stage determined it was 20 m (65 ft) above the landing area designated by Terrain-Relative Navigation, it began the Sky Crane Maneuver. Nylon cords spooled out in order to lower the rover to a position 7.6 m (25 ft) beneath the descent stage. When the spacecraft sensed the rover had touched down, pyrotechnically activated blades severed the cords and the now surplus descent stage flew a safe distance away before falling to the Martian surface.

Confirmation of landing

The NASA Mars Reconnaissance Orbiter (MRO) was passing overhead during Perseverance's landing. It received telemetry from the lander and then relayed it to Earth, allowing JPL mission controllers to confirm the touchdown. MRO had been configured to send telemetry to Earth throughout the Perseverance landing timeline in 5-second packets with a latency of about 16 seconds.

The Deep Space Network (DSN) antenna complex near Madrid, Spain was the lead station during entry, descent, and landing, with the complex in Goldstone, California providing support.

Previous antennas were limited in the frequency bands they could receive and transmit, often being restricted to communicating only with specific spacecraft. Deep Space Station (DSS)-56 in Spain was the first to employ the full range of communication frequencies. This made DSS-56 an "all-in-one" antenna able to communicate with all the missions the DSN supports and serve as a backup for

Fig. 2.6 Madrid's DSN complex. Photo courtesy of NASA/JPL-CalTech

Fig. 2.7 DSN's new dish in Madrid. Photo courtesy of NASA/JPL-CalTech

any of the Madrid complex's other antennas. With the addition of DSS-56 and other 34 m (112 ft) antennas to all three DSN complexes around the world, the network was being prepared to provide critical communication and navigation support for upcoming robotic Moon and Mars missions and the crewed Artemis missions.

Mars 2020 also provided "tones" in the X-band and ultrahigh frequency (UHF) band directly to Earth as it entered the Martian atmosphere. These dial-tone-like signals did not carry detailed engineering data but they enabled the engineers to determine that the rover had achieved key milestones and provided reassurance that its systems were still functioning. The X-band tones were received by DSN antennas. The UHF signal was received by the Green Bank Observatory in West Virginia and the Effelsberg Observatory in Germany.

Because Perseverance landed at a time of day when Jezero Crater did not have a direct line of sight back to Earth, both of these direct-to-Earth transmissions were expected to terminate a little after back shell separation, about one minute before touchdown. Backup plans scheduled MRO to replay the detailed engineering data 10 and 15 minutes after landing for additional opportunities to study the situation. NASA's Mars Atmosphere and Volatile EvolutioN (MAVEN) orbiter was also in view by when Perseverance landed and it received the same detailed engineering data as MRO. However, MAVEN was not set up to immediately transmit data to Earth immediately, so it recorded and replayed it later.

For a 3:25 minute NASA video of the Perseverance Descent and Landing go to: https://www.youtube.com/watch?v=4czjS9h4Fpg

2.3.6 Surface Operations

The surface operations phase is the time when the Perseverance rover conducts its scientific studies on Mars. After landing safely on Thursday, February 18, 2021 it initiated its primary mission plan of at least one Martian year.

One of Perseverance's first tasks in surface mode was to take a pair of pictures with the engineering imagers known as the Hazard Cameras, or HazCams, that were mounted on the front and rear of the rover. HazCams had clear covers over their lenses to protect them from the dust that was kicked up during landing. The first two images – one from the front and one from the rear – were taken through these covers mere minutes after landing. The reduced-resolution forms of these images, known as "thumbnails," were available the same day.

For a 1 hr 10 min video of the first briefing after the Perseverance landed on February 18, 2021, go to: https://www.youtube.com/watch?v=9OCxouQGnns

Over the next couple days and through the weekend (18th-21st), the rover took additional pictures of the landing site and its own hardware, including using the Navigation Cameras (NavCams) and Mastcam-Z on the remote sensing mast (the "head" of the rover). The rover also transmitted additional images from the EDL

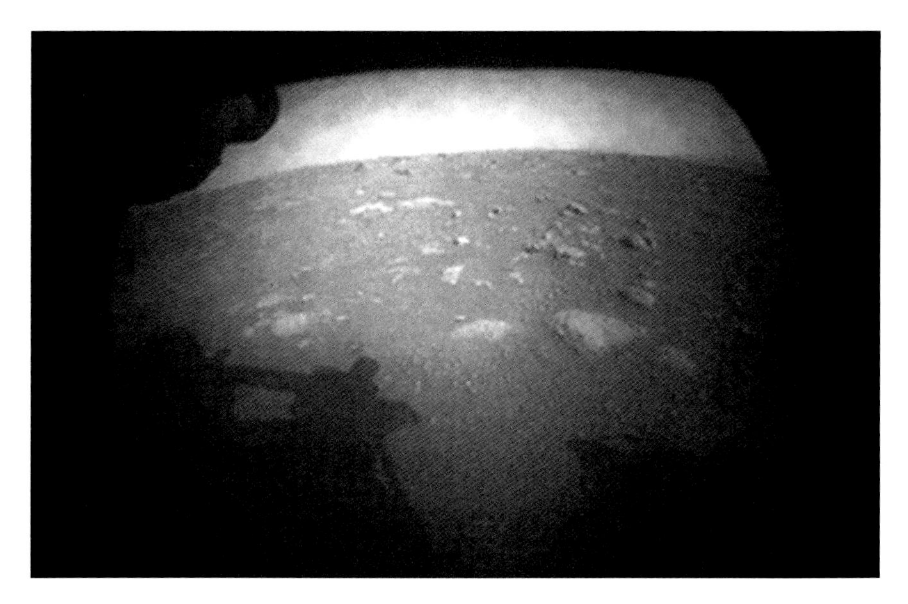

Fig. 2.8 Perseverance's first image. Photo courtesy of NASA/JPL-CalTech

cameras and microphone, allowing the public to see and hear what it was like to land on Mars.

To listen to the winds of Mars, go to: https://www.youtube.com/watch?v=FGDmZJyibYo

Checkout

Perseverance's first images were part of a planned 90-sol initial checkout period. The mission team performed tests of all the rover's parts and science instruments to ensure that everything – including the team – was ready for surface operations. For about 90 sols, the operations team operated on Mars time, meaning they set their clocks to the Martian day to allow them to respond quickly to any issue the rover might have during its workday and to ensure that revised instructions were ready for the next sol. Working on Mars time also meant the team shunted their start time 40 minutes later each time. Eventually, team members woke up in the middle of the night to start their shifts. Because living on Mars time makes daily life on Earth very challenging, the mission plan required the team to do this only for a limited period.

During its surface activities Perseverance was to seek rocks that were formed in, or were altered by, environments that may have supported microbial life in the ancient past (Objective A), find rocks capable of preserving chemical traces of ancient life (biosignatures), if any existed (Objective B), drill out core samples from about thirty promising rock and soil (regolith) targets, then cache them on the surface (Objective C), and test the extraction of oxygen (O_2) from the carbon dioxide (CO_2) atmosphere in support of future human missions (Objective D).

Perseverance will conduct the following steps during surface operations:

Step 1: Find compelling rocks

As the Perseverance rover explores Mars, scientists will identify promising rock targets. They are especially looking for rocks that formed in, or were altered by, water. This focus is because water is key to "life as we know it." Such rocks are even more interesting if they have organic molecules, the carbon-based chemical building blocks of life. Some special types of rocks can preserve chemical traces of life over billions of years. These should improve the chances of finding traces of ancient microbial life, if that ever existed. In addition, Perseverance will also collect volcanic and other rocks in order to help establish a record of geological and environmental changes over time.

Step 2: Collect rock samples

Once scientists identify a rock target of interest, Perseverance will drill a core sample. Using a pre-cleaned tube for the sample, the rover's rotary percussive coring drill penetrates about 5 cm (2 in) into the target material.

Step 3: Seal the rock samples

When finished, Perseverance will break off the core sample from the rock and cap and hermetically seal it inside the tube. Such samples will weigh about 15 gm (0.5 oz) each.

Step 4: Carry the samples

The Perseverance rover will place each sealed tube in a storage rack onboard and transport it until the mission team decides to deposit it on the Martian surface. A strategy known as Depot Caching will determine when and where to leave tubes. In the baseline plan, the rover will place one or more large groups of samples in strategic locations.

Step 5: Cache the samples

The Perseverance rover will put the Martian samples, the "witness blanks," and the "procedural blanks" in the same place on the Martian surface so that a future mission could potentially retrieve and return them all together. The mission may cache over thirty selected rock and soil samples.

Follow-on steps potentially completed by a later Mars mission

The sample cache(s) will remain on the Martian surface awaiting potential pick-up by a future mission. Images taken by orbiters will identify the locations of the samples to a precision of about 1 m (~3 ft). Images taken by Perseverance's own cameras will increase that precision to less than 1 cm (~ 0.5 in).

Almost every sol since Perseverance landed, we've watched as the team executed the daily cycle: acquire data from the rover, interpret the science data to develop a plan for the next sol's activities and observations, and implement those activities into code and beam them up to the rover at the end of its workday. This is done under enormous time pressure and with extraordinary care to eliminate mistakes that could bring the mission to a premature end. Perseverance is a technological marvel but it is the highly choreographed human processes at JPL that operate it.

IMAGE LINKS

Fig. 2.1 https://mars.nasa.gov/system/resources/detail_files/22530_PIA23302-web.jpg

Fig. 2.2 https://blogs.nasa.gov/kennedy/wp-content/uploads/sites/246/2020/08/Blast-off-Mars-2020-080520-1024x682.jpg

Fig. 2.3 https://mars.nasa.gov/system/resources/detail_files/25156_Mars_Perseverance_Trajectory_0817.jpg

Fig. 2.4 https://mars.nasa.gov/internal_resources/647

Fig. 2.5 https://mars.nasa.gov/system/resources/detail_files/25606_PIA23970-dot-1200.jpg

Fig. 2.6 https://mars.nasa.gov/system/resources/detail_files/25590_E_dsn-madrid-web.jpg

Fig. 2.7 https://d2pn8kiwq2w21t.cloudfront.net/images/jpegPIA24163.width-1600.jpg

Fig. 2.8 https://cdn.mos.cms.futurecdn.net/RWjqwbaMWkafWHzbViAuj.jpg

3

Perseverance's Design

3.1 EVOLUTION

The design of the Perseverance rover evolved from its predecessor, the Curiosity rover of the Mars Science Laboratory mission which was launched on November 26, 2011 and arrived on Mars on August 6, 2012 after a daring landing sequence that NASA called "Seven Minutes of Terror." The two rovers, both built by JPL, share many design features but play different roles in the ongoing exploration of Mars and the search for evidence of ancient life. They also differ in key mission-specific ways: while Curiosity has identified and continues to study environments where microbes could have lived early in Mars' history, Perseverance's task is to seek direct signs of such life and prepare geological samples that a future mission could bring back to Earth.

3.1.1 Curiosity

1 Curiosity has seven cameras on its mast, two of which are color cameras forming the "Mast Camera" or Mastcam. They have taken stunning color panoramas of the Martian landscape. The rover's arm extends 2 m (6.5 ft) and wields a rotating 30 kg (66 lb) turret equipped with a science camera, chemical analyzer, and rock drill.
2 Curiosity collects rock samples and studies them in its onboard laboratory. It is currently ascending the 5 km (16,404 ft) tall Mount Sharp, which may be the eroded remains of sediments that were laid down when Gale Crater held a lake.
3 Curiosity's key discoveries include uncovering evidence of ancient lakes on Mars and, with the help of a drill on its robotic arm, finding chemicals necessary for life, most notably sulfur, nitrogen, oxygen, phosphorus, and carbon.

M. von Ehrenfried, *Perseverance and the Mars 2020 Mission*,
Springer Praxis Books, https://doi.org/10.1007/978-3-030-92118-7_3

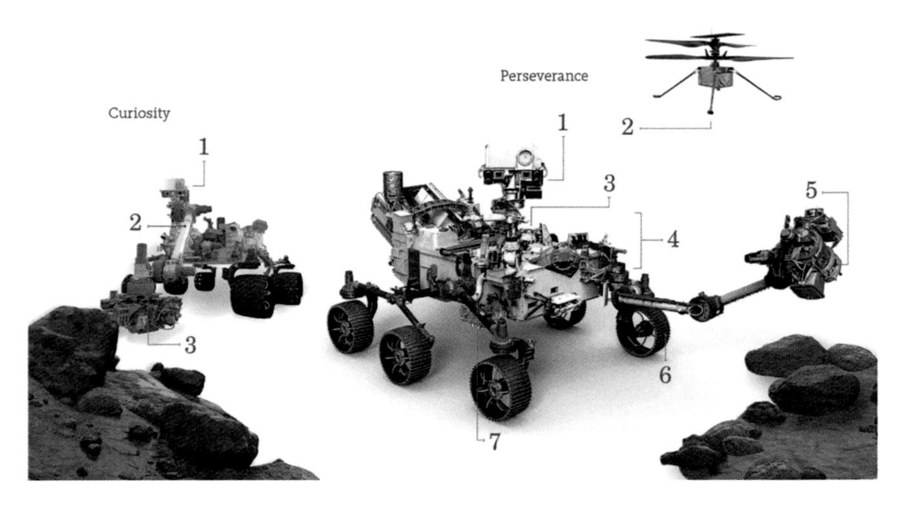

Fig. 3.1 Curiosity versus Perseverance. Photo courtesy of NASA/JPL-CalTech

3.1.2 Perseverance

1 Perseverance has the most advanced "eyes" ever sent to Mars. It has a total of 23 cameras, most of them color. It also has two microphones to capture (for the first time) the sounds of a Mars landing and the Martian wind. Mastcam-Z, an improved version of Curiosity's mast camera, has a zoom capability and can take high-definition videos and panoramas, and improved three-dimensional imagery.

2 Perseverance landed on Mars with a small helicopter drone affixed to its belly. During surface operations, the helicopter would be released to test (for the first time) powered flight on another planet.

3 An enhanced brain permits Perseverance to figure out its route on Mars autonomously, up to five times faster than was possible for Curiosity, to reduce the amount of planning time needed for navigation and allow the new rover to cover more ground and accomplish more tasks.

4 The chassis, or body, is about 12 cm (4.72 in) longer than Curiosity's. It is also heavier at 1,025 kg (2,259.7 lb), compared with 899 kg (1982 lb) for Curiosity.

5 Perseverance's arm has the same reach as Curiosity's, but its turret, at 50 kg (110 lb), weighs more because it carries larger instruments and a larger drill for coring in search of past life. The drill will cut intact rock cores in the Jezero Crater, the site of a former lake that left distinctive depositional formations, which could then be retrieved by future missions and returned to Earth for detailed study.

6 The wheels are made from the same material as Curiosity's but are bigger and narrower, with thicker skins. Instead of Curiosity's chevron-patterned treads, Perseverance has straighter ones and twice as many per wheel. The test in the

simulated Martian landscape of JPL's Mars Yard showed these treads would better withstand the pressure of driving over sharp rocks and also maneuver well on sand.

7 MOXIE, an oxygen-generating instrument the size of a car battery, will demonstrate how future explorers could extract oxygen from the Martian atmosphere.

3.2 COMPONENTS

The Perseverance rover can be thought of as having the following functional components, similar to a human:

Body:	A structure that protects the "vital organs" of the rover.
Brains:	Computers to process information.
Temperature controls:	Internal heaters, a layer of insulation, and more.
Neck and head:	A mast for the cameras to give the rover a human-scale view.
Eyes and ears:	Cameras and instruments that give the rover information about its environment.
Arm and hand:	A way to extend its reach and collect rock samples for study.
Wheels and legs:	Parts for mobility.
Electrical power:	Batteries and power.
Communications:	Antennas for "speaking" and "listening."

3.2.1 Body

The Perseverance rover's body is called the Warm Electronics Box (WEB). Like a car body, the rover's body is a strong outer layer that protects the computer and electronics, which are basically the equivalent of the rover's brains and heart. Its task is to keep the vehicle's vital organs protected and thermally stable.

The WEB is closed on the top by a piece called the Rover Equipment Deck. This makes the rover resemble a convertible car, providing a place for the rover mast and cameras to sit out in the Martian air, taking pictures with a clear view of the terrain as the rover travels.

Specifications

Main job:	Hold and protect the computer, electronics and instrument systems.
Length:	3 m (9.84 ft).
Width:	2.7 m (8.86 ft).
Height:	2.2 m (7.22 ft).
Weight/mass:	1,025 kg (2,260 lb).

The combination of larger instruments, a new sampling and caching system, and modified wheels makes Perseverance heavier at 1,025 kg (2,260 lb) compared to 899 kg (1,982 lb) for Curiosity; a 14% increase.

Structure

The bottom and sides of the WEB comprise the frame of the chassis. The top is the rover equipment deck (its "back"). The bottom is the belly pan. The first 46 cm (1.5 ft) of the belly pan measured from the front was to be dropped soon after landing in order to expose the workspace to the atmosphere and gain more room for sample handling operations within that workspace for the new Sampling and Caching system.

Fig. 3.2 Perseverance's body. Photo courtesy of NASA/JPL-CalTech

3.2.2 Brains

The JPL team created new software to run the rover. The software can be updated with improvements throughout the mission. In addition to managing the sampling tasks, Perseverance manages all of its daily activities more efficiently to balance its on-site science measurements while also collecting samples for possible future analysis. To do that, the rover's driving software, the "brains" for moving around, was modified to give Perseverance greater independence than Curiosity ever had.

The rover's brains are in its boxy body. The computer module is called the Rover Compute Element (RCE). Perseverance has two RCEs, one of which is normally asleep. In case of problems, the other RCE can be awakened to take over control and continue the mission. The RCE interfaces with the engineering functions of the rover over two networks that use an aerospace industry standard meant for the high-reliability needs of aircraft and spacecraft. In addition, they direct interfaces with all of the rover's instruments for exchange of commands and scientific data.

Specifications

The computer consists of equipment comparable to a high-end laptop. It contains special memory to tolerate the extreme radiation environment from space and to safeguard against power-off cycles, so the programs and data won't accidentally erase when the rover shuts down at night. The onboard memory includes 128 MB of dynamic random-access memory (DRAM) with error detection and correction and 3 MB of electrically erasable programmable read-only memory (EEPROM). The rover's computer uses the BAE Systems RAD750 radiation-hardened single board computer, based on a ruggedized PowerPC G3 microprocessor (PowerPC 750). The computer runs at 133 MHz. The flight software, written in C, runs on the VxWorks Operating System. It employs 4 GB of NAND (a logic gate which produces an output which is false only if all its inputs are true); thus its output is complement to that of an AND gate.

Perseverance carries an Inertial Measurement Unit (IMU) which provides 3-axis information on its position to enable the rover to make precise vertical, horizontal and side-to-side (yaw) movements. The device is used in rover navigation in order to support safe traverses and to estimate the degree of tilt the rover is experiencing on the Martian surface, permitting it to cover more ground without consulting JPL controllers so frequently. In addition, engineers have added a "simple planner" to the flight software. This allows more effective and autonomous use of electrical power and other rover resources. In particular, it allows the rover to shift the time of some activities to take advantage of openings in the daily operations schedule.

3.2.3 "Eyes" and Other "Senses"

The Perseverance rover has several cameras focused on engineering and science tasks. Some helped the landing on Mars. Others serve as the scientists "eyes" on the surface while driving around. And there are other cameras to make scientific observations and assist in the collection of samples.

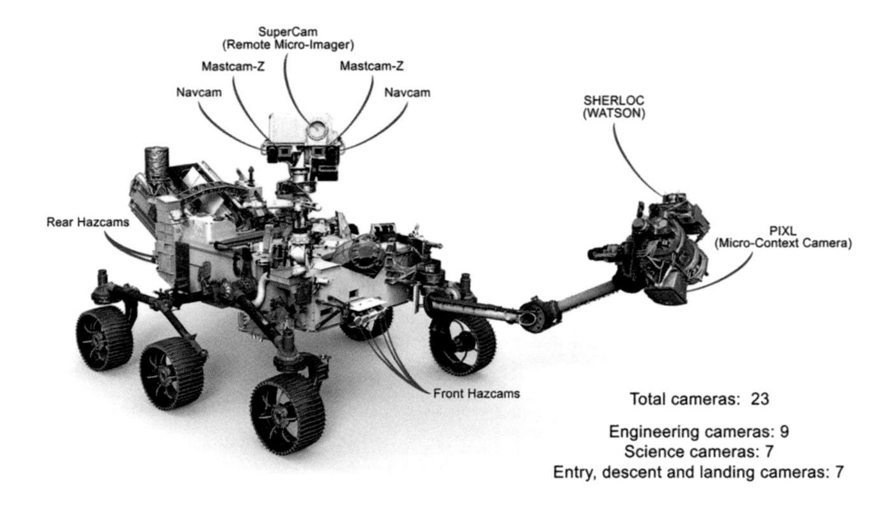

Fig. 3.3 Perseverance's cameras. Photo courtesy of NASA/JPL-CalTech

The camera suite has nine Engineering Cameras, seven Science Cameras, and seven Entry, Descent, and Landing Cameras:

Engineering cameras

These include the Hazard Avoidance Cameras (HazCams), Navigation Cameras (NavCams) and the CacheCam.

Perseverance uses a new generation of engineering cameras that build upon the capabilities of earlier Mars rovers. These "enhanced" engineering cameras give much more detailed information, in color, about the terrain in the vicinity of the rover. They have various functions: measuring the ground around the rover for safe driving, monitoring the status of rover's hardware, and supporting sample-gathering. Some help to identify the best way to move closer to tricky scientific targets.

The enhanced engineering cameras for driving help JPL operators drive the rover more precisely, and better target the movements of the arm, drill, and other tools that must be maneuvered close to their targets. A much wider field of view gives the cameras a much better view of the rover itself. This is important for checking on the health of various rover parts and measuring changes in the amount of dust and sand that may accumulate on upper surfaces. The new cameras can also take pictures while the rover is moving. The enhanced engineering cameras share the same camera body, but use different lenses selected for each camera's particular task.

Two color stereo Navigation Cameras (NavCams) help JPL engineers to navigate Perseverance safely, particularly when the rover operates autonomously, making its own navigation decisions without consulting Earth. Located high on the mast, these cameras help engineers in driving the rover. They can resolve an object as small as a golf ball from 25 m (82 ft) away. Before Perseverance "drives blind," the navigation cameras initially help ensure a safe path. The blind-drive mode is used when engineers command the rover to drive a certain distance in a specific direction, with the rover's "brains" calculating the distance traveled from wheel rotations without taking into account possible wheel slippage.

Science cameras

These include the Mastcam-Z, SuperCam, PIXL, SHERLOC and WATSON.

Mastcam-Z is a pair of cameras that provide color images and video, and three-dimensional stereo images. It also includes a zoom function. Like the Mastcam cameras on the Curiosity rover, Mastcam-Z on Perseverance consists of two duplicate camera systems mounted on the mast that stands above the rover deck. The cameras are horizontally separated and point in the same direction to give a 3-D perspective similar to what a human would see, but better.

SuperCam fires a laser at mineral targets that are beyond the reach of the rover's robotic arm and analyzes the vaporized rock to reveal its elemental composition. Like the ChemCam on Curiosity, SuperCam fires laser pulses at targets smaller than 1 mm from a distance of more than 7 m (20 ft). When the laser hits the rock, it produces plasma, which is an extremely hot gas made of free-floating ions and electrons. A spectrograph records the spectrum of the plasma in order to identify the composition of the material. It seeks organic compounds that could be related to past life on Mars.

PIXL uses X-ray fluorescence to identify chemical elements in targets as small as a grain of table salt. It has a Micro-Context Camera to provide images to correlate maps of elemental composition with the visual characteristics of the target.

In addition to a laser and spectrometers, SHERLOC uses an integrated "context" macro camera to take extreme close-ups of the areas under study. This provides context for what the laser was targeting and also helps scientists see textures that might tell the story of the environment in which the rock formed.

The WATSON camera is one of the tools on the "hand" (also called the turret) at the end of Perseverance's robotic arm. It is almost identical to the MAHLI hand-lens camera used by Curiosity. WATSON (Wide Angle Topographic Sensor for Operations and eNgineering) captures the images that bridge the scale from the very detailed images and maps that SHERLOC collects of Martian minerals and organics to the broader scales that SuperCam and Mastcam-Z see from atop the mast. WATSON provides views of the very-fine-scale textures and structures in

Martian rocks and the rocky debris and dust that covers so much of the Martian surface. These capabilities allow WATSON not only to support SHERLOC but also to help to identify targets of interest for the other rover instruments. Since WATSON can be moved around by the robotic arm, it also provides images of instruments and rover parts. For example, it can be aimed at the oxygen-making experiment MOXIE to help monitor how much dust accumulates at the inlet that lets in Martian air for the extraction of oxygen. A target which includes a metric standardized bar graphic is mounted on the front of the rover's body to calibrate WATSON.

Entry, descent, and landing cameras

The Mars 2020 Entry, Descent, and Landing Camera suite included:

- Parachute "up look" cameras mounted on the back shell in order to view the deployment and inflation of the parachute. Two of the three cameras successfully recorded the chute.
- Descent-stage "down look" camera mounted on the descent stage to view the lowering of the rover during the sky crane maneuver.
- Rover "up look" camera mounted on the top deck of the rover to view the descent stage during the sky crane maneuver and descent stage separation.
- Rover "down look" camera mounted beneath the rover to view the surface during landing.

In addition to providing valuable engineering data, the cameras can be considered a "public engagement payload." They certainly gave a dramatic sense of the ride down to the surface! Aside from computer animations, there had never been any views of a parachute opening in the Martian atmosphere, a rover being lowered to the surface of Mars, or the descent stage flying away after rover touchdown. With Perseverance, we had a "front row seat" to a Mars landing for the first time in the history of space exploration.

In particular, these videos provided data to enable the team explore:

- How does the parachute deploy and operate in the Martian atmosphere?
- How does the landing system move as it descends and nears the surface?
- How much sand and rock is stirred up by the retro rockets?
- What exactly happens when the rover touches down?
- Precisely where did the vehicle touch down in the landing area?
- How can an aerial perspective obtained shortly prior to landing inform plans for driving the rover?

Over the years, technology had replicated much of the human sensory experience on Mars. Cameras gave us sight, robotic hands, arms and feet supplied touch, and chemical and mineral sensors enabled us to taste and smell on Mars. Hearing was the last of the five senses to be exercised on the Red Planet.

Recording the sounds on Mars

Two earlier NASA spacecraft to Mars have carried microphones. Unfortunately, one of those missions, the Mars Polar Lander, failed. The Phoenix Lander had a microphone incorporated into the descent camera, but that instrument was never turned on.

Perseverance was equipped with two microphones to finally enable us to listen to the sounds during entry, descent and landing on Mars. The "brains" of the system were inside the body of the rover and the "ears" were on side to record the sounds of descent, friction from the atmosphere, and dust blown up by the thrusters as the rover descends

The SuperCam toolkit also included a microphone to help to study rocks and soil during surface operations. This would enable scientists to hear the staccato "pop" caused by the laser zapping a rock, wind, and rover noises. Located on a 15 mm (0.6 in) boom on the head of the rover's mast it can listen for a few milliseconds at a time when SuperCam is active, or to listen to the wind and rover sounds for about 3.5 minutes at a time. This microphone was adapted from easily available, store-bought hardware and weighs 30 gm (~1 oz).

For a 3:09 video of some of the sounds recorded by SuperCam, go to:
https://www.youtube.com/watch?v=EiQvlM4taBQ

3.2.4 Wheels and Legs

The Perseverance rover has six wheels, each with its own motor. The two front and two rear wheels also have individual steering motors to allow the vehicle to swerve and curve, make arcing turns, and even turn a full 360 degrees in place.

Wheels specifications

The wheels are made of aluminum with cleats for traction and curved titanium spokes for springy support. They are 52.5 cm (20.7 in) in diameter, therefore a full turn of the wheels with no slippage drives the rover 1.65 m (65 in).

Perseverance uses a similar "rocker-bogie" suspension system that was also used on the Mars Science Laboratory, Mars Exploration Rovers, and Pathfinder rover missions. The suspension system connects the wheels to the rest of the rover and controls how the rover interacts with the Martian terrain. The suspension system has three main components:

- Differential: This connects to the left and right rockers and to the body by a pivot in the center of the rover's top deck.
- Rocker: One each on the left and right side of the rover, connect the front wheel to the differential and the bogie in the rear.
- Bogie: Connects the middle and rear wheels to the rocker.

Fig. 3.4 Perseverance's wheels. Photo courtesy of NASA/JPL-CalTech

The suspension system maintains a relatively constant pressure on every wheel when driving over the uneven terrain and also minimizes vehicle tilt as it drives, thereby enhancing its stability.

Fig. 3.5 Perseverance maneuvers to check out its drive system. Photo courtesy of NASA/JPL-CalTech

The rocker-bogie suspension enables the rover to drive over obstacles (such as rocks) or through depressions that are as large as the wheel. Each wheel has an aggressive tread composed of 48 grousers (or cleats) machined into its surface which give the vehicle excellent traction when driving in both soft sand and on hard rocks.

Perseverance is designed to endure a tilt of 45 degrees in any direction without tipping over, but to minimize risk the drivers at JPL avoid situations that would cause a tilt of more than 30 degrees.

Driving speed

By Earth-vehicle standards, the Perseverance rover is slow. By Martian-vehicle standards it is a standout performer able to cross flat, hard ground at 4.2 cm/sec (152 m/hr), or a little less than 0.1 mi/hr. For comparison, a 3 mph walking pace is 134 cm/sec (4,828 m/hr). In the case of exploring Mars, however, speed isn't the most important quality. It's about the journey and the destinations along the way. Perseverance's pace is energy efficient, consuming less than 200 watts of power.

Legs

The legs are made of titanium tubing formed with the same process that is used to make high-end mountain bike frames, and they enable Perseverance to drive over knee-high rocks as tall as 40 cm (15.75 in).

3.2.5 Arm and Turret

The 2.1 m (7 ft) long robotic arm on Perseverance can move a lot like a human arm, with shoulder, elbow and wrist "joints" for maximum flexibility. It lets the rover work as a human geologist would, by holding and using science tools with its "hand" or turret. The rover's own "hand tools" extract cores from rocks, take microscope images, and analyze the elemental and mineral makeup of rocks and soil.

The arm uses tiny motors called "rotary actuators" to achieve five degrees of freedom known as the shoulder azimuth joint, shoulder elevation joint, elbow joint, wrist joint and turret joint.

At the end of the arm is the "turret" that serves as the rover's hand by carrying scientific cameras, mineral and chemical analyzers for studying rocks and soils, and choosing the most scientifically valuable samples to cache.

The turret holds the Gaseous Dust Removal Tool (GDRT), the SHERLOC and WATSON and PIXL instruments, the Ground Contact Sensor, and the Drilling and Coring mechanism.

Fig. 3.6 Perseverance's arm. Photo courtesy of NASA/JPL-CalTech

Fig. 3.7 The turret. In this image taken on July 11, 2019, engineers at JPL install the sensor-filled turret onto the end of the Perseverance rover's 2.1 m (7 ft) long robotic arm. Photo courtesy of NASA/JPL-CalTech

SHERLOC

The Scanning Habitable Environments with Raman & Luminescence for Organics & Chemicals (SHERLOC) instrument is intended to study minerals up close, so it is mounted on the turret where it can be placed next to its targets. SHERLOC uses spectrometers, a laser, and a camera to search for organics and minerals that have been altered by watery environments and may be signs of past microbial life. See also Section 3.4.

WATSON

Perseverance's Wide Angle Topographic Sensor for Operations and eNgineering (WATSON) camera is like a geologist's hand-lens for magnifying and recording the textures of rock and soil targets studied by the SHERLOC mineral analyzer. It is also assists PIXL, and can provide valuable views of rover systems such as the wheels and instruments low on the rover out the field of view of Mastcam-Z. See also Section 3.4.

PIXL

The Planetary Instrument for X-ray Lithochemistry (PIXL) carried on the turret is capable of detecting changes in textures and chemicals in Martian rocks and soil left behind by any ancient microbial life. The information it supplies will be used to decide which Martian samples are the most scientifically interesting candidates for caching. See also Section 3.4.

Ground contact sensor

The turret includes a sensor to signal the arm to stop if it inadvertently comes into contact with the ground in order to prevent damage.

Drill and coring mechanism

The turret has a drill that can employ rotary motion with or without percussion to penetrate rock and is equipped with three different kinds of attachments (bits) that facilitate sample acquisition and surface analysis. The coring and regolith bits are used to collect Martian samples directly into a clean sample collection tube, while another bit is used to scrape or "abrade" the outer layers of rocks to expose fresh, unweathered surfaces for examination.

The cylindrical drill cuts out samples from rock interiors, breaking off the rock core at its base. Each sample is collected directly into a clean tube that is about the size of a penlight. A core of rock is 13 mm (0.5 in) in diameter and 60 mm (2.4 in) long, and typically 10–15 gm of Martian material. A special drill bit is used to collect the loose soil that seems to be ubiquitous on the Martian surface. As with samples of rocks, these samples of regolith are collected directly into a clean sample collection tube.

3.2.6 Sample Handling

Perseverance is the first Mars rover capable of taking samples of rocks and soil and store them in tubes that can be cached for later retrieval by another mission.

The three major steps in sample handling are:

Step 1: Collecting samples

A key task for the rover is collecting carefully selected samples of rock and soil that will be sealed in tubes and cached in a well-identified location (or possibly more than one location) on the surface of Mars. Detailed maps will be provided for any future mission that might retrieve the samples for study by scientists in terrestrial laboratories.

The belly of the rover houses all the apparatus and supplies necessary to collect samples. There is a rotating carousel, essentially a wheel that contains different kinds of drill bits. Next to it are the 43 sample tubes, waiting to be filled. While the rover's arm reaches out and drills rock, the rover belly is home to a smaller robotic arm that works as a "lab assistant" to the main arm. The small arm picks up and feeds new sample tubes to the drill and coring mechanism, and transfers the filled containers to a space where they are sealed and stored.

Perseverance must meet extraordinary cleanliness requirements. These measures are in place to avoid contaminating Martian samples with anything that may have been inadvertently brought from Earth. Strict rules limit the amount of inorganic, organic, and biological materials from Earth in the rover and its sample handling system.

Perseverance carries five "witness tubes" along with the sample collection tubes. The witness tubes are similar to the sample tubes except they are pre-loaded with a variety of materials which can capture molecular and particulate contaminants such as gases that could be released or "outgassed" by different materials on the rover, chemicals from the firing of the landing propulsion system, and any other terrestrial organic or inorganic materials that may have arrived on Mars with the rover.

One at a time, the witness tubes will be opened on the Martian surface in order to "witness" the environment in the vicinity of sample collection sites. After being exposed to the local environment they will be put through the motions of drilling and other actions that the sample containers experience. Although these tubes do not collect soil or rock samples, they will be sealed and cached together with the actual samples.

In the future, if the Perseverance samples are returned to Earth for analysis, the witness tubes will show whether Earth contaminants were present during sample collection. This will help scientists to tell which materials in the samples may be of terrestrial origin.

Step 2: Storing onboard

After a sample has been collected, its tube is transferred back to the rover's belly where it is handed off to the small interior robotic arm, then moved to inspection and sealing stations. Once a tube has been hermetically sealed, nothing can enter or leave it. The tubes will be stored in the belly until the team decides on a place to cache the samples.

Step 3: Depositing samples on the surface

At a time of the team's choosing, the samples will be deposited on the surface at a spot designated as a "sample cache depot." The depot location (or locations) must be well-documented by both local landmarks and precise coordinates from orbital measurements. The cache of Mars samples will remain at the depot, available for retrieval by a future mission and return to Earth.

See Chapter 5 for the sample caching that was undertaken during Perseverance's first campaign.

3.2.7 Communications

Perseverance has three antennas located on the equipment deck (its "back") that serve as both its "voice" and its "ears." Having a number of antennas provides operational flexibility as well as back-up options. The antennas on rover's deck include:

- Ultra-high frequency antenna.
- X-band high-gain antenna.
- X-band low-gain antenna.

Ultra-high frequency antenna

Most often, Perseverance will use its ultra-high frequency (UHF) antenna (about 400 MHz) to communicate with Earth through NASA spacecraft in orbit around Mars. Being within close range of each other, the antennas of the rover and the orbiters act like walky-talkies in contrast to the long-range telecommunications with Earth provided by the low-gain and high-gain antennas.

It generally takes about 5 to 20 minutes for a radio signal to travel the distance between Mars and Earth at the speed of light, depending on planetary positions. Relaying a message via an orbiter is beneficial because an orbiter is much closer to Perseverance than the Deep Space Network on Earth. The mass- and power-constrained rover can achieve high data rates of up to 2 Mbps on the relatively short-distance relay link to an orbiter. The orbiter then employs its much larger antenna and more powerful transmitter to relay the signal across interplanetary space to Earth.

The X-band high-gain antenna

The high-gain antenna located mid-aft on the portside of the deck is steerable to enable it to point its radio beam in a specific direction. It is hexagonally shaped, 0.3 m (1 ft) in diameter and operates at 7–8 GHz.

The benefit of having a steerable antenna is that the entire rover does not need to change position in order to talk to Earth, which is always moving in the Martian sky. Like turning your neck to talk to someone standing beside you is preferable to turning your entire body, the rover can save energy and keep things simple by moving only the antenna. Its high gain allows it to focus its beam for higher data rates on the long link back to Earth via the NASA Deep Space Network.

The X-band low-gain antenna

Perseverance employs its low-gain antenna primarily for receiving signals. This antenna can send and receive information in every direction; that is, it is omnidirectional. Because it does not need to be aimed, it provides a robust means of communicating with the rover. It transmits at a low data rate to the Deep Space Network. It can receive signals sent using the 34 m (112 ft) diameter antennas at approximately 10 bps or faster or from the 70 m (230 ft) diameter antennas at 30 bps or faster.

3.2.8 Energy Source

Radioisotope Thermoelectric Generators (RTGs) provide electrical power for a spacecraft by converting the heat generated by the decay of plutonium-238 (Pu-238) fuel into electricity using devices called thermocouples. Since they have no moving parts that can fail or wear out, RTGs have historically been viewed as a highly reliable power option. Thermocouples have been used in RTGs for a total combined time exceeding 300 years without a single thermocouple ever ceasing to produce power.

Thermocouples are found in everyday items that must monitor or regulate their temperature, such as air conditioners, refrigerators, and medical thermometers. The principle of a thermocouple involves two plates, each made from a different metal that conducts electricity. Joining the two plates to form a closed electrical circuit while the two junctions are at different temperatures produces an electric current. Each of these pairs of junctions forms an individual thermocouple. In an RTG the radioisotope fuel heats one junction while the other junction is exposed to the space environment or a planetary atmosphere.

RTGs have been powering spacecraft, landers and rovers for over 50 years. The current model used by NASA is the Multi-Mission Radioisotope Thermoelectric Generator (MMRTG). The one on Perseverance has a mass of 45 kg (99 lb) and holds 4.8 kg (11 lb) of Pu-238 oxide. The heat liberated by radioactive decay is converted by thermocouples to electricity. At the time of mission launch it was delivering approximately 110 watts but this will diminish in line with Pu-238's half-life of 87.7 years.

The MMRTG charges two lithium-ion batteries that power the rover's activities and must be recharged periodically. Unlike solar panels, the MMRTG provides engineers with significant flexibility in operating the rover's instruments even at night, during dust storms, and through winter.

Fig. 3.8 The Multi-Mission Radioisotope Thermoelectric Generator. Photo courtesy of NASA/JPL-CalTech

3.2.9 Markings

From bracelets to body art, humans have adorned themselves for thousands of years. The spacecraft we send to Mars are no different! Many NASA orbiters, landers and rovers fly with artwork, signs, and symbols that reflect where and when they were created. As you might expect, Perseverance has the American flag, logos of NASA, JPL, Mars 2020, plus a tattoo, a logo representing health workers and a rover evolution plate. But it also has some other embellishments etched onto pieces of titanium or aluminum, some of which celebrate previous missions while others offer hope for future human achievements on Mars.

Fig. 3.9 The calibration target. Photo courtesy of NASA/JPL-CalTech

The calibration target for the SHERLOC instrument carries some hidden gems. The bottom row includes spacesuit materials to see how they react over time to the radiation in the Martian atmosphere. One is a sample of polycarbonate that could be used for a helmet visor. It doubles as a geocaching target and is etched with 221B Baker Street, the address of the beloved fictional detective Sherlock Holmes and his chronicler Dr. John Watson.

Fig. 3.10 The coded parachute was imaged by a camera on the protective back shell during the entry, descent and landing phase of the Mars 2020 mission. Photo courtesy of NASA/JPL-CalTech

A distinctive pattern was incorporated into the white and orange sections of the 70-foot-wide supersonic parachute for the Mars 2020 mission. Upward-looking cameras on the back shell took images of the parachute while descending in the Martian atmosphere to help engineers to determine the precise orientation of the parachute as it inflated. (Remember that this was the first time that a lander was able to observe the deployment of its parachute.) Engineers seized on this as an opportunity to make a binary brain puzzle. Within each circular row of the chute they added the words "Dare Mighty Things" using binary code. The phrase was taken from a speech by President Theodore Roosevelt. More pertinently it is the JPL motto. Along the outer edge of the chute are the Global Positioning System coordinates for JPL. There are now T-shirts, hoodies, and even dresses with this image.

Three small squares on a placard on the rover contain silicon chips that hold the names of 10,932,295 people who participated in the "Send Your Name to Mars" campaign, as well as 155 essays submitted by students who entered a contest to name the vehicle.

Fig. 3.11 NASA/JPL-CalTech provided the author's ticket to Mars.

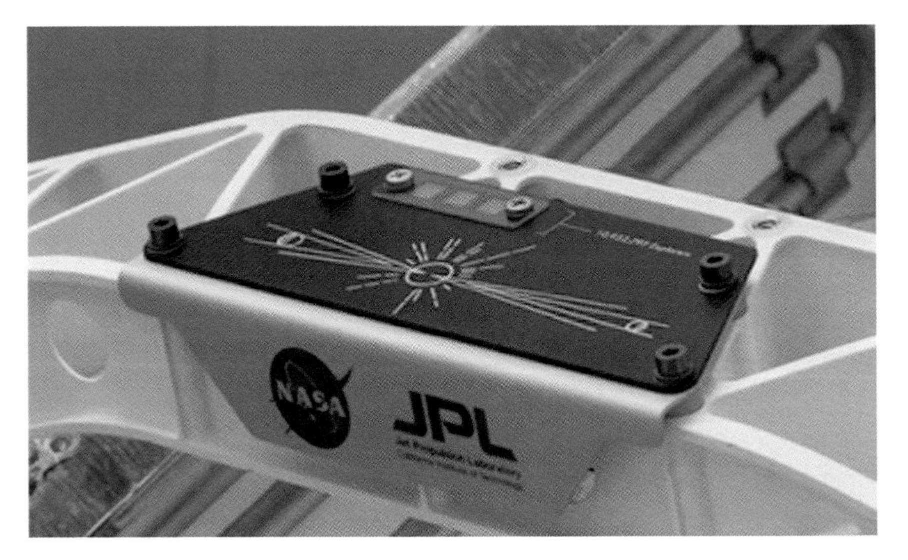

Fig. 3.12 The three small squares on the upper left of this placard on the rover contain a variety of messages from Earth. Photo courtesy of NASA/JPL-CalTech

3.3 STAGES

The Mars 2020 spacecraft consists of several mechanical components: the cruise stage, descent stage, back shell, and heat shield plus the rover. These major parts of the design were based on the successful flight of the Mars Science Laboratory which delivered the Curiosity rover to Mars in August 2012.

After its 7-month interplanetary cruise, the spacecraft had to maneuver through the Martian atmosphere to deposit Perseverance on the surface. In the process of entry, descent and landing, certain parts of the spacecraft were discarded, one by one, until the rover was safely on the ground.

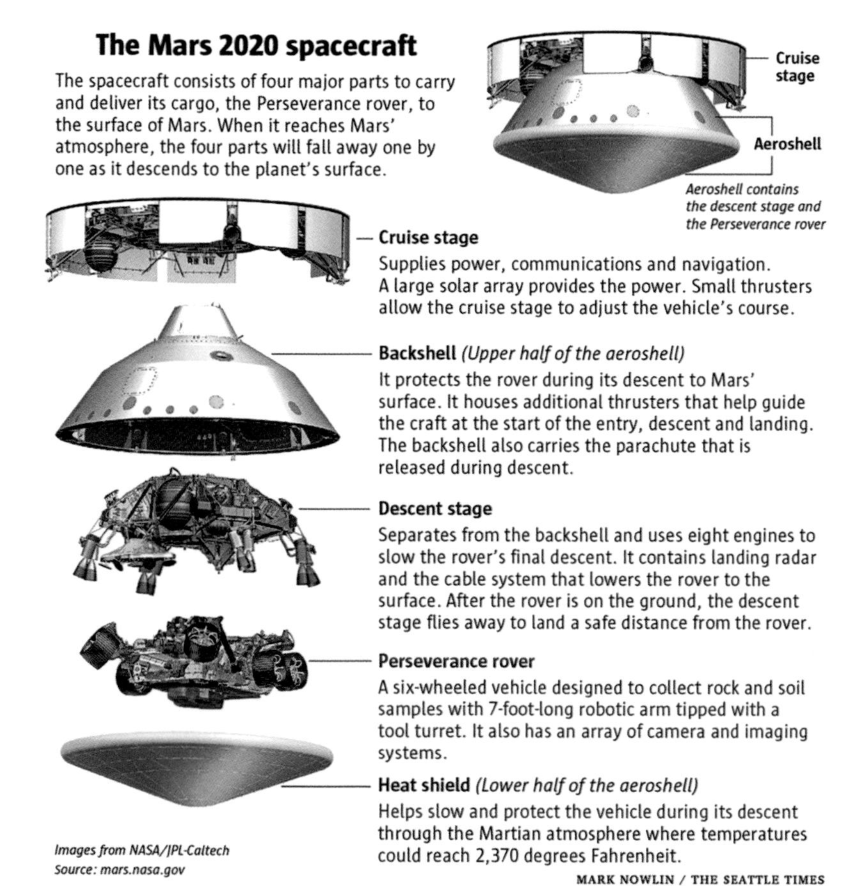

Fig. 3.13 Mars 2020 spacecraft expanded view. Photo courtesy of the Seattle Times/ NASA/JPL-CalTech

3.3.1 Cruise Stage

JPL engineers kept close tabs on the mission during the cruise to Mars, with the major activities including:

- Checking spacecraft health and maintenance.
- Monitoring and calibrating the spacecraft and its onboard subsystems and instruments.
- Performing attitude correction turns in order to keep the antenna aimed at Earth for communications and the solar panels facing the Sun for power generation.

- Conducting navigation activities to determine and correct the flight path and train navigators before atmospheric entry.
- Preparing communications systems to be used during the entry, descent, and landing (EDL) phase and subsequent surface operations.

The cruise stage supported the whole vehicle during the lengthy flight to Mars, keeping it powered up, in communication, and on trajectory. A large solar array provided power. Radio antennas maintained contact with Earth. Fuel tanks and small thrusters allowed it to adjust the trajectory as necessary.

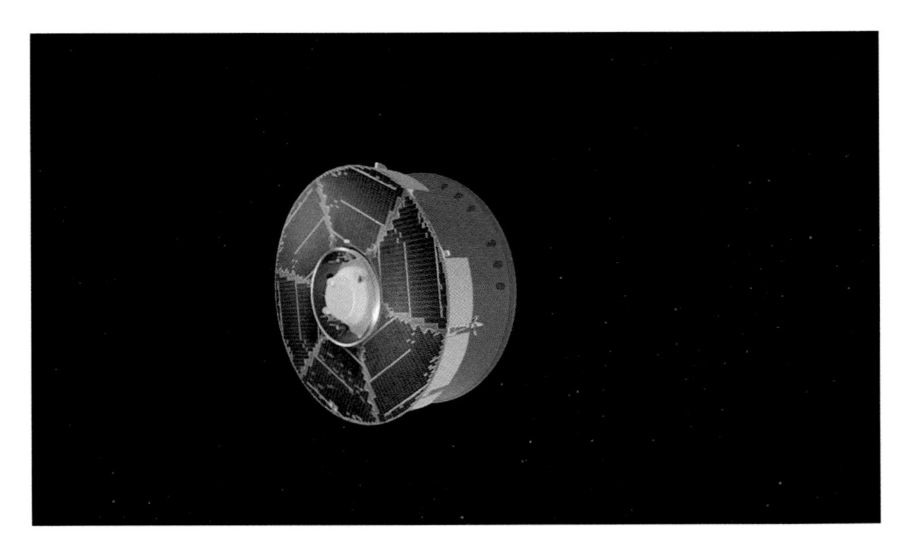

Fig. 3.14 The Mars 2020 cruise stage. Photo courtesy of NASA/JPL-CalTech

Approximately 2.65 m (8.7 ft) in diameter and 1.6 m (5.2 ft) tall when mounted on the aeroshell, the cruise stage was similar to that of Mars Pathfinder. It had a launch mass of 1,063 kg (2,344 lb). Its primary structure was aluminum with an outer ring of ribs that were covered by five sections of solar arrays which would provide up to 600 watts of power near Earth and 300 watts at Mars.

Small heaters and multi-layer insulation blankets kept the electronics "warm" in space but there was a Freon system to remove heat from the flight computer and telecommunications hardware inside the rover to prevent them from overheating. Cruise avionics systems allowed the flight computer in the rover to interface with other systems such as the Sun sensors and the star scanner (with a backup). From the positions of the Sun and other stars in relation to itself the spacecraft was able to determine where it was in space.

Fig. 3.15 The Perseverance rover casting off its cruise stage, minutes before entering the Martian atmosphere. Photo courtesy of NASA/JPL-CalTech

About 10 minutes prior to atmospheric entry the cruise stage was to separate from the aeroshell, enclosing the rover and its descent stage. The aeroshell would make the trip to the surface on its own.

A spacecraft on a 515 million km (320 million mi) flight to Mars can sometimes be slightly off course, so the plan for the cruise schedules a number of trajectory correction maneuvers (TCMs).

During the Mars 2020 cruise phase there were five nominal opportunities (plus one backup maneuver and one contingency maneuver) to adjust the flight path. For each TCM, engineers calculated the direction in which the spacecraft must point the eight thrusters on the cruise stage and the length of time they would have to fire in order to tweak the path to ensure the spacecraft would enter the Martian atmosphere at just the right spot for a landing inside Jezero Crater.

The final 45 days of the cruise known as the approach phase mainly involved navigation activities and preparing for the Entry, Descent and Landing. There were opportunities for up to three final trajectory correction maneuvers.

To ensure the spacecraft arrived at Mars in the right place for its planned landing two lightweight aluminum-lined tanks carried a maximum capacity of 31 kg (68 lb) of hydrazine propellant. Together with cruise guidance and control systems, these tanks of propellant allowed navigators to maintain the spacecraft

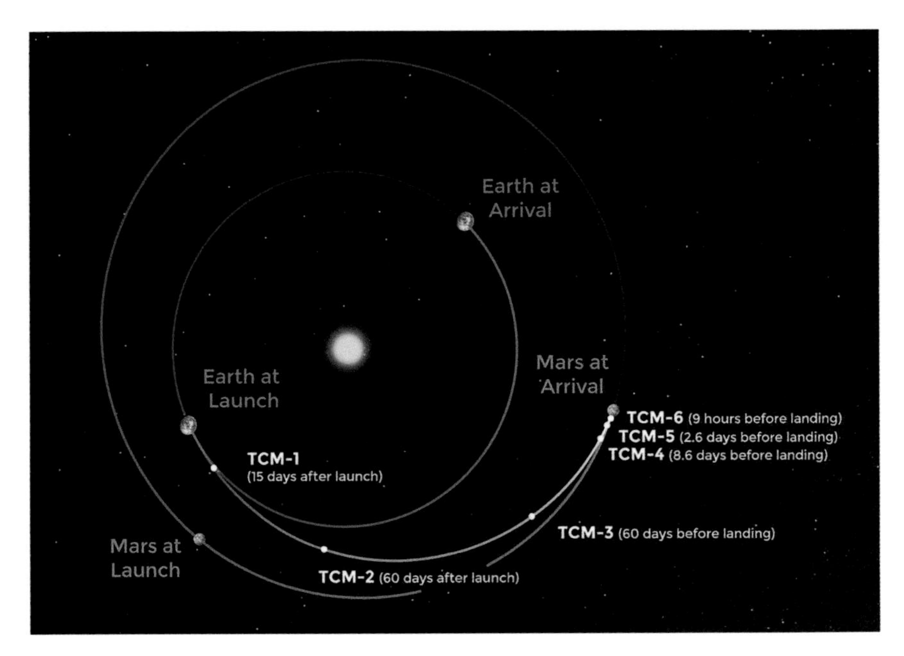

Fig. 3.16 The trajectory to Mars. Photo courtesy of NASA/JPL-CalTech

closely on course. By burns and pulse firings, the propellant enabled three different types of trajectory correction maneuvers:

- An axial burn using pairs of thrusters to change spacecraft's velocity.
- A lateral burn using two "thruster clusters" (four thrusters per cluster) to move the spacecraft "sideways" through seconds-long pulses.
- Pulse mode firing using coupled thruster pairs for spacecraft precession maneuvers.

JPL navigators would send the commands through two X-band antennas on the cruise stage:

- The Cruise Low-gain Antenna was mounted inside the inner ring of the cruise stage. During flight the spacecraft was spin-stabilized with a spin rate of 2 rpm. Periodic spin axis pointing updates made sure the antenna kept pointing toward Earth and that the solar arrays remained facing the Sun. The low-gain antenna was used during the early part of the cruise, when the spacecraft was close to Earth.
- The Cruise Medium-Gain Antenna was mounted in the outer ring. As the spacecraft moved farther from Earth it switched over to the tighter beam of this antenna.

3.3.2 Back Shell and Rover

The back shell was to protect the rover during its turbulent descent through the Martian atmosphere. The back shell also housed additional thrusters that would fire during the "guided" portion of entry, descent and landing. Inside the top of the back shell was the canister containing the parachute.

Fig. 3.17 The back shell. Photo courtesy of NASA/JPL-CalTech

Fig. 3.18 The Perseverance rover in its folded configuration. Photo courtesy of NASA/JPL-CalTech

Fig. 3.19 The stack of three components. Mated to its interplanetary cruise stage, the Perseverance rover sits within a bell-shaped back shell shortly prior to being attached to the brass-colored heat shield at the Kennedy Space Center. The next time they separated would be at an altitude of 10 km (6.2 m) above Jezero Crater on February 18, 2021. Photo courtesy of NASA/JPL-CalTech/KSC

3.3.3 Heat Shield

The heat shield helps to slow the vehicle down during its entry into the Martian atmosphere and protect the rover inside from the intense heat experienced. The heat shield was made by Lockheed Martin out of phenolic-impregnated carbon ablator (PICA) invented by NASA's Ames Research Center. The heat shield on the Mars 2020 spacecraft reached approximately 1,300°C (2,370°F) traveling at more than 19,550 kph (12,500 mph). At an altitude of about 8 km (5 mi) above the surface the vehicle deployed its parachute and the heat shield "popped off."

Fig. 3.20 The heat shield mated to the back shell. Both components were 4.5 m (14.8 ft) in diameter. The aeroshell encapsulated and protected the descent stage containing the rover both during the interplanetary cruise and the penetration of the Martian atmosphere. The image was taken at Lockheed Martin Space in Denver, Colorado, which manufactured the aeroshell. Photo courtesy of NASA/Kim Shiflett

3.3.4 Descent Stage

As the rover's free-flying "jetpack," the descent stage separated from the back shell and used eight engines to slow the final descent. It also carried the landing radar system needed to make last-minute decisions about touchdown. This phase of the EDL sequence was called "powered descent." Hundreds of critical events would have to be executed perfectly and exactly on time in order for the rover to land safely on February 18, 2021.

At an altitude of about 2,100 m (6,900 ft) the vehicle shed the back shell and its parachute and the descent stage fired up its engines and flew to a reachable self-selected safe landing target, leveled out, and slowed to its final descent speed of about 2.7 kph (1.7 mph).

Fig. 3.21 The descent stage separation test. Photo courtesy of NASA/JPL-CalTech

Fig. 3.22 As the descent stage nears the surface it fires up its engines. Photo courtesy of NASA/JPL-CalTech

Fig. 3.23 The sky crane maneuver. Photo courtesy of NASA/JPL-CalTech

The sky crane maneuver was initiated with about 12 seconds of flight remaining, roughly 20 m (66 ft) above the surface. The descent stage lowered the rover on a set of cables 6.4 m (21 ft) long. Meanwhile the rover itself unstowed its mobility system, locking its legs and wheels into landing position. Once the rover was on the ground, the cables were severed and the descent stage flew away to make an uncontrolled landing on the surface a safe distance away from the rover.

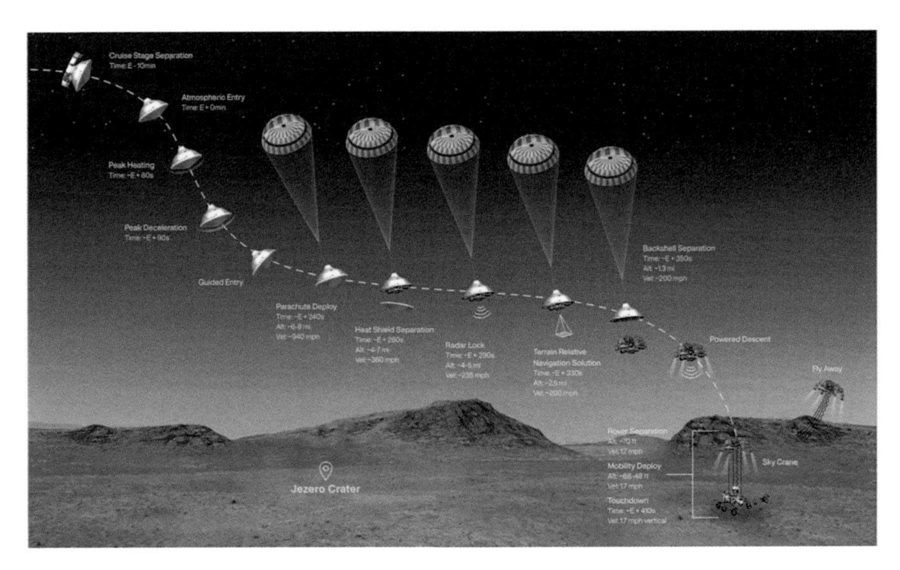

Fig. 3.24 The entry sequence of events. Photo courtesy of NASA/JPL-CalTech

For a 3:10 minute animation of the landing, go to:
https://www.youtube.com/watch?v=rzmd7RouGrM

Fly alongside Perseverance in this 3-D demo of its Entry, Descent, and Landing:
https://mars.nasa.gov/mars2020/timeline/landing/entry-descent-landing/

For an 8:08 minute video of the Perseverance Entry, Descent and Landing, go to:
https://www.youtube.com/watch?v=aNTPcFbixPg

3.4 INSTRUMENTS

3.4.1 Background

Following the established processes of NASA's Science Mission Directorate, the instrument selection process began with the establishment of a Science Definition Team to outline the mission's objectives, realistic surface operations, a proof-of-concept instrument suite and suggestions for threshold science measurements that would meet the proposed objectives.

Once NASA accepted and/or modified these recommendations, it released an Announcement of Opportunity on July 1, 2013. Scientists and engineers on competitive teams focused on designing instruments to match the established criteria, then NASA selected optimal components from among the submitted proposals.

In order to answer compelling research questions, NASA openly competed the opportunity for the mission's specific science payloads and then announced the selection of the Mars 2020 rover science instruments on July 13, 2014. Science instruments would support studies related to habitability, the search for potential signs of past microbial life, identifying the most compelling samples of rock and soil for future potential return to Earth, and preparing for eventual future human exploration.

The Perseverance rover was to carry seven primary instruments. Although these instruments were briefly described in previous chapters, the following is a more technical and scientific account.

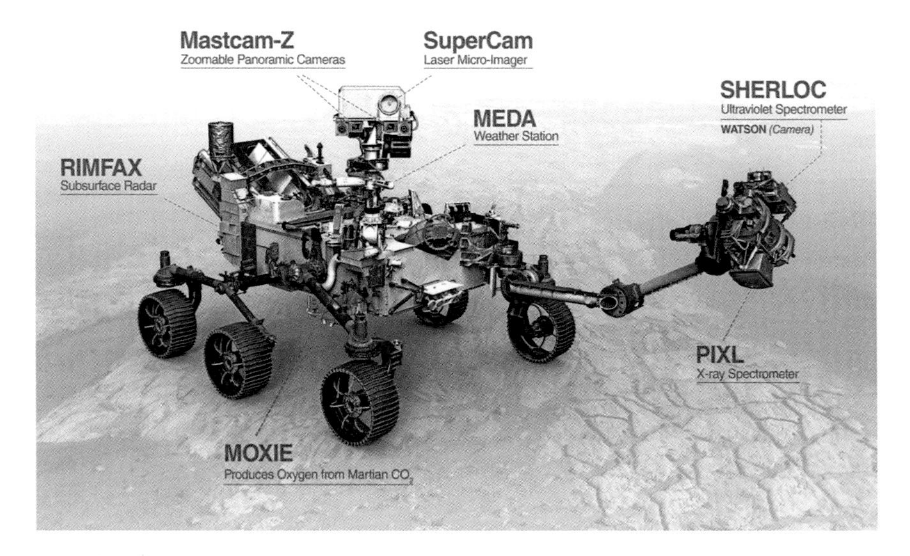

Fig. 3.25 Perseverance's instruments. Photo courtesy of NASA/JPL-CalTech

3.4.2 Mastcam-Z

This instrument is a multispectral, stereoscopic imager based on the successful Mastcam on the Curiosity rover. From its position on the Remote Sensing Mast of Perseverance, Mastcam-Z acquires visible color (RGB), stereo panoramas of the Martian surface at resolutions sufficient to resolve ~1 mm details in the near-field (i.e. the workspace of the robotic arm) and ~3–4 cm features 100 m (328 ft) away. It also has bandpass filters (400–1000 nm) that can distinguish unweathered from weathered materials and yield insights into the mineralogy of many silicates, oxides, oxyhydroxides, and diagnostic hydrated minerals. Mastcam-Z also images the Sun directly using a pair of solar filters.

The Mastcam-Z cameras have the capability to zoom, focus, acquire data at high-speed (video rates of 4 frames per second or even faster for subframes) and hold large numbers of images in internal storage. These capabilities allow investigators to examine targets that are otherwise beyond the rover's reach. The cameras can observe time-dependent phenomena such as dust devils, motions of clouds, and astronomical phenomena as well as activities involved in driving, sampling and caching operations, and the flights of the Ingenuity helicopter. Mastcam-Z has improved stereo imaging capabilities compared to Curiosity's Mastcam and the Pancam of the Spirit and Opportunity Mars Exploration Rovers. It also provides advances in navigational and instrument-placement capabilities that can help to support and enhance Perseverance's driving and coring/sampling capabilities.

Mastcam-Z observes textural, mineralogical, structural, and morphologic details in rocks and soils at the rover's field site. The cameras allow the science team to piece together the geological history of the site by way of the stratigraphy of rock outcrops and the regolith, as well as to constrain the types of rocks present, such as sedimentary versus igneous types.

The instrument has three overarching science roles:

- Characterize the overall landscape geomorphology, processes, and the nature of the geological record (i.e. mineralogy, texture, structure, and stratigraphy) at the rover field site.
  - Instrument observations provide a full description of the topography, geomorphology, geological setting and the nature of past and present geological processes of the Perseverance field site, especially as they pertain to habitability.
  - Studies include observations of rocks and outcrops to help determine morphology, texture, structure, mineralogy, stratigraphy, rock type, history/sequencing, and the associated depositional, diagenetic, and weathering characteristics.
  - Characterizing the overall landscape and its geological record also requires observations of regolith to help evaluate the physical and chemical alteration, together with stratigraphy, texture, mineralogy, and depositional/erosional processes.
- Assess current atmospheric and astronomical conditions, events, and surface-atmosphere interactions and processes.
  - Mastcam-Z makes observations of clouds, dust-raising events, the properties of suspended aerosols (dust, ice crystals), astronomical phenomena, and aeolian (wind) transport of fine-grained particles.
  - Mastcam-Z images can characterize potential ice- or frost-related (periglacial) geomorphic features, and even the characterization of frost or ice, if present, and its influence on rocks and fines.

- Provide operational support and scientific context for rover navigation, contact science, sample selection, extraction and caching, and the other selected Perseverance investigations as well as the Ingenuity helicopter.
  - Mastcam-Z images assist with rover navigation by determining the location of the Sun and of horizon features, and giving information pertinent to rover traversability (such as distant hazards and terrain meshes).
  - Mastcam-Z observations enable other Perseverance instruments to identify and characterize materials to be collected for in-situ study, coring and caching, or other purposes (including the monitoring of hardware).

Fig. 3.26 The twin Mastcam-Z cameras shown with a pocket knife for scale. Photo courtesy of Malin Space Science Systems/ASU

The Principal Investigator for Mastcam-Z is Dr. Jim Bell from Arizona State University.

3.4.3 MEDA

This simple abbreviation for the Mars Environmental Dynamics Analyzer spells two Spanish words that can be roughly translated as "give me," as in, "MEDA! Give me the weather, dust and radiation report on Mars!"

This suite of sensors can record the optical properties of dust and six atmospheric parameters: wind speed and direction, pressure, relative humidity, air temperature, ground temperature, and radiation in discrete bands of the ultraviolet, visible, and infrared ranges of the spectrum. The radiation sensor is part of an assembly with two arrays of photodiodes that also detect low-elevation-angle scattered light and a sky-pointing camera, and their data are combined to characterize the properties of atmospheric aerosols.

Systematic measurement is the primary driver for MEDA operations. Over the entire mission's lifetime with a configured cadence and frequency in accordance to resource availability, MEDA records data from all sensors. Implementation of this strategy is based on a high degree of autonomy in MEDA operations. It will awaken itself on an hourly basis, record and store data and then go back to sleep independently of rover operations. It records data whether the rover is awake or not, and both day and night.

The main science objectives for the science team are:

- Signature of the Martian general and mesoscale circulation on phenomena near the surface (e.g. fronts, jets).
- Microscale weather systems (e.g. boundary layer turbulence, heat fluxes, eddies, dust devils).
- Local hydrological cycle (e.g. spatial and temporal variability, diffusive transport from regolith).
- Dust optical properties, photolysis rates, ozone (O_3) column, and oxidant production.

As an environmental instrument, MEDA's different sensors are in direct contact with ambient conditions:

- The radiation/dust sensor assembly is located on top of the rover's deck.
- The pressure sensor is inside the body and connected to the atmosphere through a dedicated pipe.
- All other sensors are located around the Remote Sensing Mast (RSM):
    - Two wind sensor booms oriented at 120 degrees from each other measure winds approaching the RSM.
    - Five sets of thermocouples measure the air temperature.

- A thermal infrared sensor measures downward and upward thermal infrared radiation as well as surface skin temperature.
- A relative humidity sensor is also attached to the RSM.

The full suite of sensors is controlled by the instrument control unit within the rover body.

The sensors that characterize the Martian low atmosphere and the properties of dust are as follows:

Air temp: Located around the RSM are three air temperature sensors. There are two more on the rover's body to ensure that one of them will be upwind. Placed on small thermal inertia forks, and outside the rover's thermal boundary layers, these five sets of three thermocouples measure atmospheric temperature across the range 150–300 K with a required accuracy of 5 K and a resolution of 0.1 K. For example in August 2021, which was late spring on Mars, the high was 0°F and the low was –115°F.

Humidity: Located at the RSM inside a protecting cylinder, a humidity sensor measures the relative humidity with an accuracy of 10% across the 200–323 K range with a resolution of 1%. A filter on the cylinder protects the sensor from dust deposition.

Pressure: Located inside the rover body and connected to the external atmosphere through a tube another sensor collects pressure measurements. The tube exits via a small opening which is protected against dust deposition. Its measurement range goes from 1–1,150 Pa with an end-of-life accuracy of 20 Pa (the calibration tests gave values around 3 Pa) and a resolution of 0.5 Pa. Because this component has to be in contact with the atmosphere, a HEPA filter is placed on the tube inlet to avoid contaminating the Martian environment.

Radiation and dust: Located on the rover deck, the radiation and dust sensor has eight upward-looking photodiodes in the following ranges:

- 255 ± 5 nm for the O_3 Hartley Band center.
- 295 ± 5 nm for the O_3 Hartley Band edge.
- 250–400 nm for total ultraviolet.
- 450 ± 40 nm for Mastcam-Z cross-calibration.
- 650 ± 25 nm for SuperCam cross-calibration.
- 880 ± 5 nm for Mastcam-Z cross-calibration.
- 950 ± 50 nm for near infrared.
- one panchromatic (at least 300–1,000 nm) filter with an accuracy better than 8% of the full range for each channel, computed based upon Mars' radiation levels and minimum dust opacity.

The photodiodes face the zenith direction and have a field of view of 30 degrees except for the panchromatic one that has a 180 degree field of view. An array of side-looking photodiodes at 880 nm characterize the low angle light scattering at various azimuth angles. They cover at least 270 degrees of a circle with fields of view of ±15 degrees and separated about 45 degrees from each other.

A dedicated camera with ±60 degrees around the zenith measures the intensity of the solar aureole.

The assembly of photodiodes and camera are located on the rover's deck without dust protection. To mitigate dust degradation magnetic rings were placed around the photodiodes in order to maximize their operational lifetimes. Nevertheless to evaluate degradation due to dust deposition, images of the sensors will be taken periodically and their readings compared with opacities from other Perseverance optical instruments such as Mastcam-Z. Comparing the differences in estimated sky opacities by MEDA and Mastcam-Z will provide evaluations of the level of dust deposition.

Thermal radiation: Attached to the RSM and pointing to the front right side of the rover, the thermal infrared sensor measures the net infrared thermal radiation near the surface of Mars with a set of five thermopiles: three downward-pointing ones for bands 16–20, 6.5-cut-on and 8–14 microns, and two upward-pointing ones for bands 6.5-cut-on and 14.5–15.5 microns.

Wind: Two wind sensors attached to the RSM measure wind speed and direction. These magnitudes are derived from the information provided by six 2D detectors on each boom. The detectors are located on six boards, 60 degrees apart around the boom axis. Each records local speed and direction in the plane of the board. The combination of the six boards per boom serves to determine wind speed, as well as the pitch and yaw angle of each boom relative to the flow direction. The requirement is to determine horizontal wind speed with 2 m/sec accuracy in the range of 0–40 m/sec, with a resolution of 0.5 m/sec. The directional accuracy is expected to be better than 22.5 degrees. In the case of a vertical wind the range is 0–10 m/sec and the accuracy and resolution are the same as for horizontal winds.

The booms hold the detectors out of the RSM's thermal boundary layer, aiming to minimize the wind flow perturbation by the RSM at the boom tip where the these detectors are located. The two booms are separated in azimuth to help ensure that at least one of them will record clean wind data while the other is in the wake of the rover. To correct for the perturbations at the booms by the RSM and the rover on the environmental temperature and wind field, a variety of numerical analyses and wind tunnel tests were used during calibration under the conditions on Mars. Numerical simulations were used to obtain results for tests conditions that could not be reproduced on Earth.

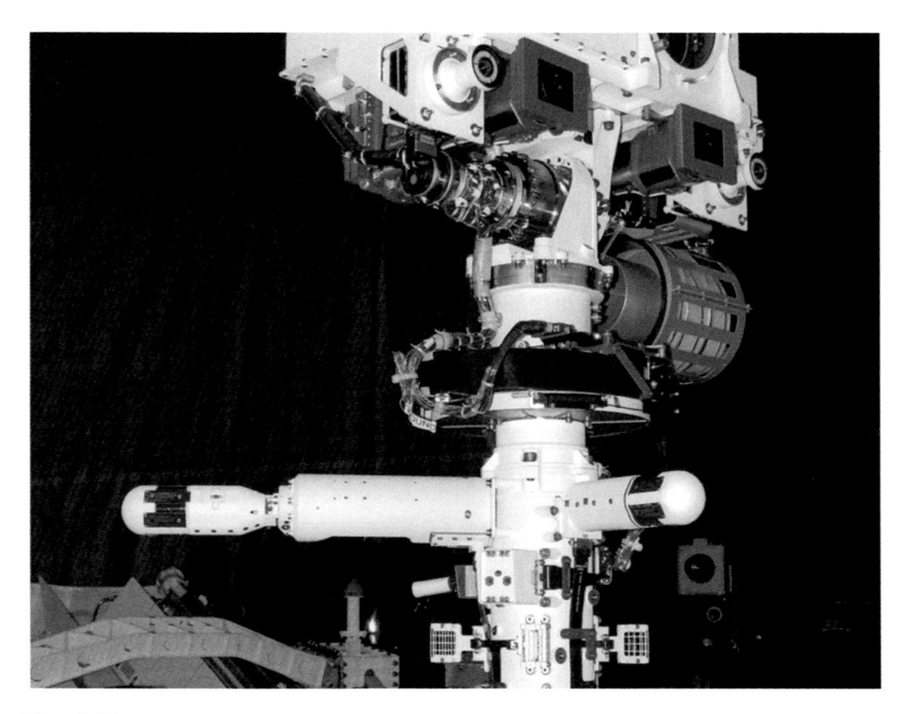

Fig. 3.27 The Mars Environmental Dynamics Analyzer (MEDA). Photo courtesy of NASA/JPL-CalTech

The Principal Investigator for MEDA is Dr. Josė A. Rodriguez Manfredi of the Centro de Astrobiologia, Instituto Nacional de Tecnica Aeroespacial in Madrid, Spain.

3.4.4 MOXIE

The Mars Oxygen ISRU eXperiment (MOXIE) investigation by the Mars 2020 mission aims to address key knowledge gaps in preparing for human exploration of the Red Planet, including demonstrating In-Situ Resource Utilization (ISRU) technologies to enable propellant and consumable oxygen production from the Martian atmosphere.

MOXIE weighs 8 kg (17.7 lb) on Earth, consumes 300 watts and is 23.9 x 23.9 x 30.9 cm (9.4 x 9.4 x 12.2 in). It draws in CO_2 from the Martian atmosphere and then electrochemically splits the CO_2 molecules into O_2 and CO. The O_2 is then analyzed for purity before being vented back to the atmosphere together with the CO and other exhaust products.

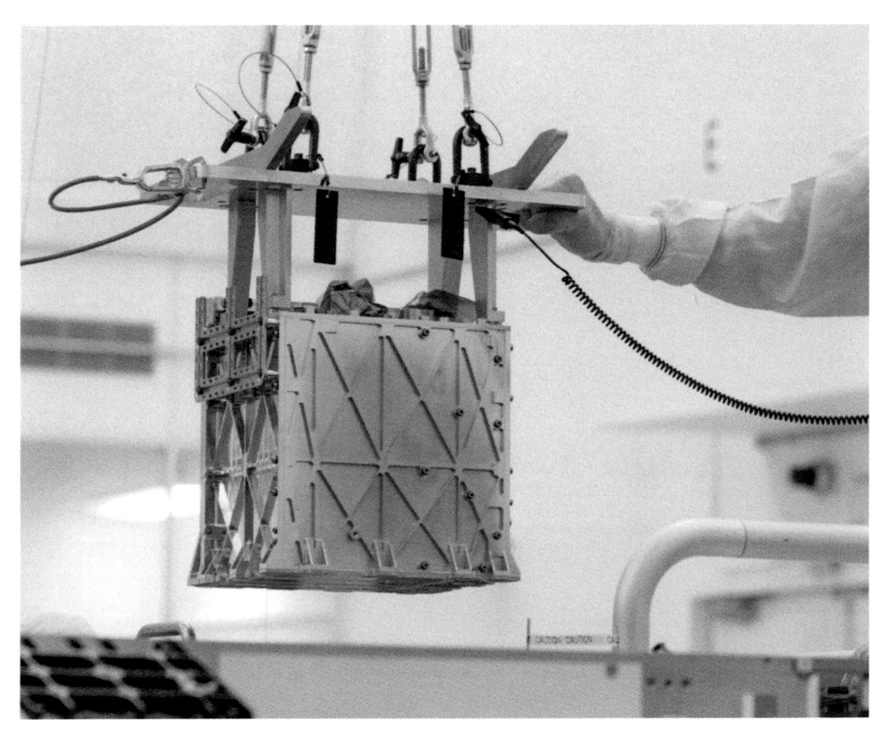

Fig. 3.28 Technicians in the clean room are lowering the MOXIE instrument into the belly of the Perseverance rover. Photo courtesy of NASA/JPL-CalTech

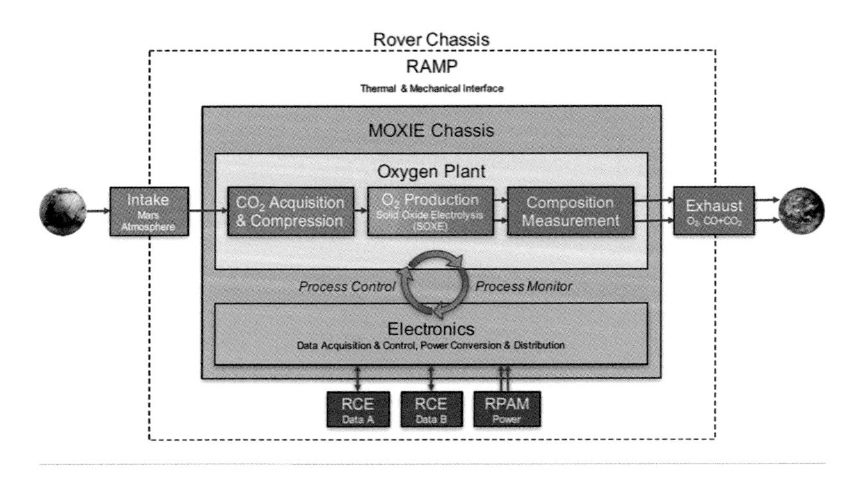

Fig. 3.29 MOXIE functional block diagram. Photo courtesy of NASA/JPL-CalTech

MOXIE's CO_2 Acquisition and Compression (CAC) system draws in Martian atmosphere through a filter and pressurizes it to approximately Earth sea level. The pressurized gas is then regulated and fed to the Solid OXide Electrolyzer (SOXE), where it is electrochemically split at the cathode to produce O_2 at the anode; a process equivalent to running a fuel cell in reverse.

Because SOXE operates at 800°C (1,472°F) it requires a sophisticated thermal isolation system, including input gas preheating and exhaust gas cooling. There are O_2 exhaust and CO_2/CO exhaust streams that are then analyzed to verify O_2 production rate and purity and for process control. The electrical current through SOXE is a direct result of the oxide ions transported across the electrolyte, so it serves as an independent measurement of the O_2 production rate. It can produce up to 10 grams per hour (at least 0.022 pounds per hour). There is approximately an hour of O_2 production per experiment, which will be scheduled intermittently over the duration of the mission.

Based upon conversion efficiency calculated from flow rates and composition measurements, the SOXE control parameters such as the CO_2 input flow rate, temperature and applied voltage can be used to optimize O_2 production under Mars environmental conditions. The cooled exhausts are then filtered to satisfy planetary protection requirements and vented. Process telemetry is reported to Perseverance for downlink.

The Principal Investigator for MOXIE is Dr. Michael Hecht of MIT.

3.4.5 PIXL

The name of the Planetary Instrument for X-ray Lithochemistry (PIXL) refers to "pixel" (the contraction of "picture element," namely the smallest digital point in an image). The pixel is at the heart of image processing and digital images, from space telescope pictures to rover "selfies." What makes PIXL special is its focus on some of the tiniest features on Mars! Along with a tip of the hat to its camera, the name also honors the "X" of its X-ray system.

The PIXL is a micro-focus X-ray fluorescence instrument that rapidly measures elemental chemistry at submillimeter scales by focusing an X-ray beam to a tiny spot on a target rock or soil and then analyzing the induced X-ray fluorescence.

Scanning the beam reveals spatial variations in chemistry in relation to fine-scale geological features such as laminae, grains, cements, veins and concretions. The high X-ray flux gives high sensitivity and short integration times: most elements are detected at lower concentrations than were possible using the instruments on previous surface missions and several new elements could be detected that were previously undetectable. Rapid acquisition allows rapid scanning, enabling PIXL to reveals the associations between different elements and the observed textures and structures. The same spectra can be summed for bulk analysis, and so allow comparison with bulk chemistry measurements at other sites already explored on Mars. With PIXL's simple design came operational efficiency and experimental flexibility. It could be adapted to different scientific opportunities, to produce a diverse set of scientifically powerful data products within the constraints of the mission.

Because PIXL works at night, it has light emitting diodes circling its opening to take pictures of rock targets in the dark. Using artificial intelligence, PIXL relies on the images to determine how far away it is from the target that it is to scan.

The instrument consists of a main electronics unit in the rover's body that weighs 2.6 kg (~6 lb). The sensor head mounted on the turret at the end of the robotic arm weighs 4.3 kg (~10 lb). The sensor head is 21.5 x 27 x 23 cm (~ 8.5 x 10.5 x 9 in). It includes an X-ray source, X-ray optics, X-ray detectors, and high voltage power supply, as well as a micro-context camera (MCC) and light-emitting diode (LED). It is able to detect elements Na, Mg, Al, Si, P, S, Cl, K, Ca, Ti, V, Cr,

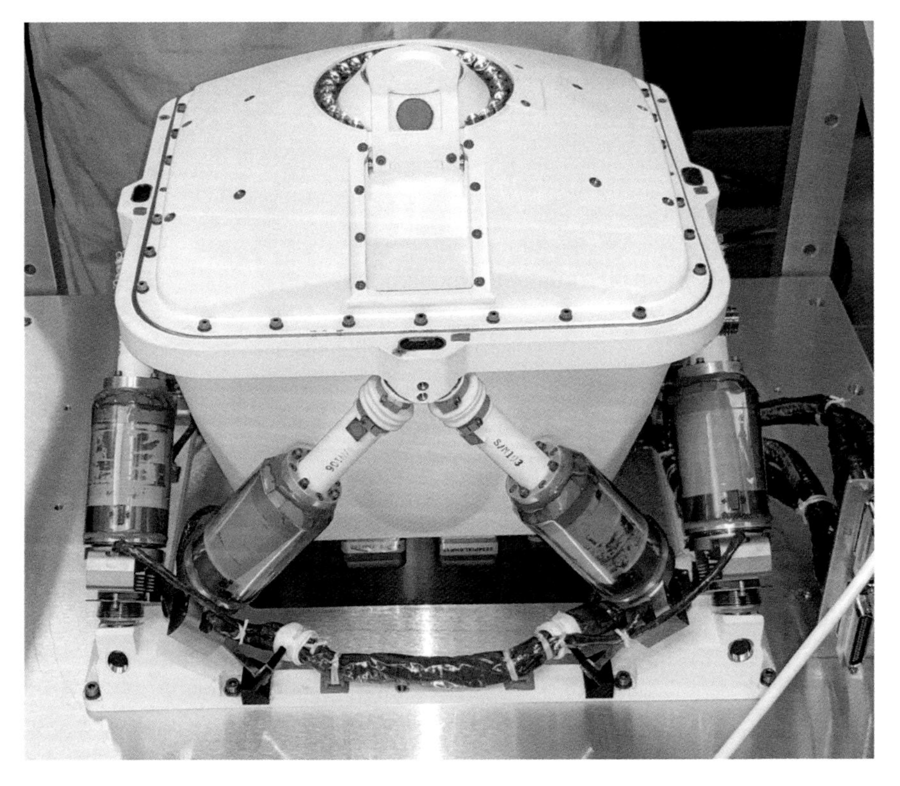

Fig. 3.30 The sensor head of PIXL prior to being mounted on the turret at the end of the robotic arm at JPL. Photo courtesy of NASA/JPL-CalTech

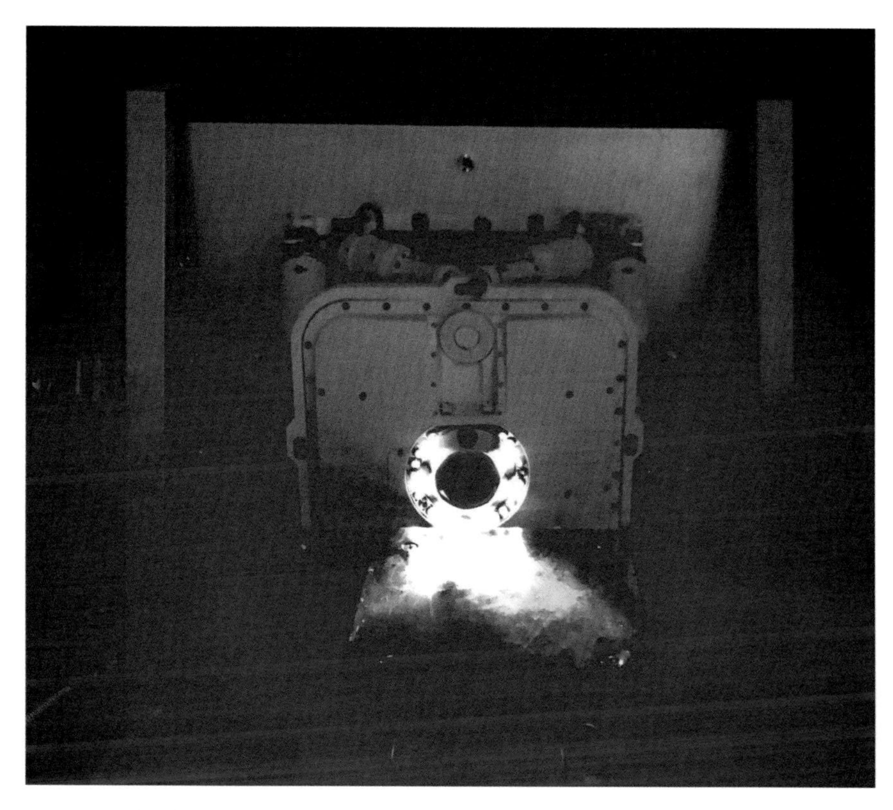

Fig. 3.31 PIXL's night lite. Photo courtesy of NASA/JPL-CalTech

Mn, Fe, Co, Ni, Cu, Zn, Br, Rb, Sr, Y, Ga, Ge, As, and Zr, with important trace elements such as Rb, Sr, Y and Zr being detectable at the level of tens of parts per million (ppm). The sensor draws about 25 watts.

The advantages of the PIXL instrument are:

- High spatial resolution with a 0.12-mm diameter beam making possible submillimeter scale geochemistry to be correlated with texture.
- Measurement flexibility: point, line, or map analysis with a variable step size as small as 0.1 mm.
- Spatial coverage, targeting and position knowledge: a fine step size and the micro-context camera provide a resolution and footprint at hand-lens scale.
- Fast spectral acquisition: measure most major and minor elements at 0.5 wt% in 5 seconds.
- Large range of detectable elements (over 26 elements).
- High sensitivity to detect important trace elements at tens of ppm level.
- Operational efficiency: The combined effect of PIXL's speed, sensitivity, resolution, coverage/targeting, and measurement flexibility enables it to

search efficiently, locate the places of interest and observe small patches of clean rock.
- Operational robustness: tolerant to surface roughness.

In order to measure the chemical makeup of rocks at a very fine scale, PIXL has four calibration targets 5 mm (0.19 in) in size weighing 0.015 kg (0.033 lb) each. The data return is around 16 megabits per experiment, or about 2 megabytes per day.

The Principal Investigator for PIXL is Dr. Abigail Allwood of JPL.

3.4.6 RIMFAX

The Radar Imager for Mars' subsurFAce eXperiment (RIMFAX) is a ground-penetrating radar (GPR) selected to fly on Perseverance and designed to obtain from Mars (for the first time by a rover) high-resolution stratigraphic evidence about the subsurface.

RIMFAX builds on mature GPR technology used on Earth for a wide variety of scientific and engineering applications adapted for use for Perseverance. Its ultra-wideband design, operating from 150 MHz to 1.2 GHz, affords a theoretical limit of 14.2 cm (6 in) for vertical (range) resolution in free-space. The instrument has an electronics box inside the rover and a downward pointing antenna mounted on the rear. It can operate as Perseverance drives and will be commanded to produce individual soundings in different modes, some for shallow and others for deeper penetration. The default operation will produce interleaved pairs of shallow- and deep-soundings at every 10 cm (3.94 in) along a traverse. The expectation is for RIMFAX signals to achieve penetrations of 10 m (32.8 ft) but it may well exceed that for subsurface conditions that are friendly to the propagation of radar waves.

The overall goal of RIMFAX is to image the subsurface structure and constrain the nature of the material which underlies the landing site. This is made possible because the propagation of radar waves is sensitive to the dielectric properties of materials, such that the variations in composition and porosity across geological strata give radar reflections that can be identified, mapped, and interpreted in the geological sense.

Specifically, RIMFAX supports and enhances the Perseverance investigation in the following (but not limited) ways:

- Assess the depth and extent of the regolith.
- Detect different subsurface layers and their relationship to visible surface outcrops.
- Characterize the stratigraphic section from which a cored-and-cached sample derives, including crosscutting relations and features which are indicative of past environments.

Finding water ice, a potential resource for drinking or making fuel, would be a valuable find for future astronauts exploring Mars.

Fig. 3.32 The RIMFAX electronics box prior to being integrated into the Perseverance rover at JPL. Photo courtesy of NASA/JPL-CalTech

Fig. 3.33 A prototype of the RIMFAX antenna attached to a snowmobile undergoing tests in Svalbard, Norway. Photo courtesy of NASA/FFI

3.4.7 SHERLOC and WATSON

Sherlock Holmes was a fictional detective who solved crimes. He used forensic methods which included scientific observation and powers of logical reasoning. SHERLOC also observes and measures. Its task is to seek indications of ancient microbial life. Dr. John Watson was Holmes' partner in solving mysteries. The WATSON camera helps SHERLOC in seeking to solve mysteries about life on Mars. SHERLOC even has a modern version of the hand-lens magnifying glass frequently used by the classic British detective!

SHERLOC is an arm-mounted Deep UltraViolet (DUV) resonance Raman and fluorescence spectrometer utilizing a 248.6-nm laser and <100 micron spot size. The laser is integrated in an autofocusing/scanning optical system, and co-boresighted to a context imager that has a spatial resolution of 30 microns. It allows non-contact, spatially resolved, and highly sensitivity detection/characterization of organics and minerals in the surface and near subsurface.

Its goals are to assess past aqueous history, detect the presence and preservation of potential biosignatures and to support selection of return samples. To do this, SHERLOC detects CHNOPS (an acronym for the six main chemical elements in organisms, namely carbon, hydrogen, nitrogen, oxygen, phosphorus and sulfur) and correlates surface texture with the distribution and type of any organics that are preserved.

SHERLOC operates over a 7 x 7 mm (0.28 x 0.28 in) area by use of an internal scanning mirror. The 500-micron depth of view and an autofocus mechanisms of MAHLI (Mars Hand Lens Imager) heritage enables it to be accurately placed 48 mm (1.89 in) above a natural or abraded surface without operating the arm's own joints. Additionally, after sample core removal, the interiors of the boreholes can be analyzed as a proxy for direct core analysis.

In addition to its combined spectroscopic and macro imaging role, SHERLOC is able to benefit from a "second-eye" in the form of the Wide Angle Topographic Sensor for Operations and eNgineering (WATSON) whose near-field-to-infinity imaging capability is used for engineering operations and science imaging. It is a build-to-print camera based on the MAHLI of the Curiosity rover. Integration is enabled by existing electronics within SHERLOC.

Deep UV-induced native fluorescence is very sensitive to condensed carbon and aromatic organics, permitting detection at or below 10^{-6} w/w mass percentage of the solute in solution (1 ppm) at less than 100 micron spatial scales. SHERLOC's deep UV resonance Raman enables detection and classification of aromatic and aliphatic organics with sensitivities of 10^{-2} to below 10^{-4} w/w at spatial scales of less than 100 micron. In addition, this method allows detection and classification of minerals relevant to aqueous chemistry with grain sizes less than 20 microns.

SHERLOC's methodology exploits two spectral phenomena: native fluorescence and pre-resonance/resonance Raman scattering. These phenomena happen when a high-radiance, narrow line-width, laser source illuminates a sample. Organics that fluoresce absorb the incident photon and then re-emit at a higher wavelength. The difference between the excitation wavelength and the emission wavelength relates to the number of electronic transitions, which increases with increasing aromatic structures (i.e. the number of rings). This phenomenon is highly efficient, with a typical cross section a hundred times greater than Raman scattering and provides a powerful means to find trace organics.

The native fluorescence emission of organics extends from about 270 nm into the visible. This is useful, because it "creates" a fluorescence-free region (from 250–270 nm) where Raman scattering can occur. With SHERLOC's narrow-linewidth 248.6 nm laser, further characterization by Raman scattering from aromatics and aliphatic organics and minerals can be observed. Furthermore, excitation using a DUV wavelength enables resonance and pre-resonance signal enhancements (in the x100 to x10,000 range) of organic/mineral vibrational bonds by coupling of the incident photon energy to the vibrational energy. This allows high-sensitivity measurements with low backgrounds without the need for a high-intensity laser, and it avoids the damage or modification of organics by inducing reactions with species such as perchlorates, which are known to exist on Mars.

The Principal Investigator for SHERLOC is Dr. Luther Beegle of JPL.

Fig. 3.34 A close-up view of an engineering model of SHERLOC. Located on the end of Perseverance's robotic arm, this instrument features an auto-focusing camera (pictured) that shoots black-and-white images which are used by the color camera called WATSON to zero in on rock textures. SHERLOC also has a laser, which aims for the dead center of rock surfaces depicted in WATSON's images. The laser uses a technique called Raman spectroscopy to detect minerals in microscopic rock features. That data is then superimposed on WATSON's images to produce "mineral maps" that help scientists to determine which samples should be drilled for cores to be sealed in metal tubes and left on the Martian surface for a future mission to retrieve and return to Earth. Photo courtesy of NASA/JPL-CalTech

3.4.8 SuperCam

SuperCam uses remote optical measurements and laser spectroscopy in order to measure fine-scale mineralogy, chemistry and atomic/molecular composition of samples encountered on Mars.

It is, in fact, many instruments in one:

- To measure elemental composition it integrates the remote laser-induced breakdown spectroscopy capabilities of the highly successful ChemCam on the Curiosity rover. It uses a 1,064-nm laser to investigate targets up to 7 m (23 ft) from Perseverance.
- In addition, SuperCam also performs Raman spectroscopy (at 532 nm) to investigate targets up to 12 m (39 ft) from the rover, and time-resolved fluorescence spectroscopy, visible and infrared reflectance spectroscopy (400–900 nm, 1.3–2.6 microns) to obtain data about the mineralogy and molecular structure of samples being considered for direct investigation, as well as being able to search directly for organics.
- Finally, SuperCam also acquires high-resolution images of samples under study using a color remote micro-imager. The data provided by this suite of correlated measurements on a sample can be used to determine directly the geochemistry and mineralogy of samples.

SuperCam measurements can be rapidly acquired without the need to position the rover or its arm on the target, facilitating rapid and efficient measurements during Mars operations. As demonstrated by ChemCam, the SuperCam laser can be used to "blast" dust off of surfaces at a distance in order to obtain a better look at solid surfaces without having to drive up and perform manipulations using the rover's arm or associated tools.

SuperCam is mounted on the "head" of the rover's long-necked mast and weighs 5.6 kg (12 lb). It is 38 x 24 x 19 cm (15 x 9 x 8 in) in size. The electronics unit is mounted in the rover's body and weighs 4.8 kg (10.6 lb). It also has a calibration target 3 cm (1.18 in) in diameter which weighs 0.2 kg (0.5 lb). The system draws 17.9 watts and has a data return of 15.5 megabits per experiment or 4.2 megabits per day.

SuperCam is a joint effort between Los Alamos and the French Space Agency (CNES), the Institut de Recherche en Astrophysique et Planétologie (IRAP) in Toulouse, the University of Hawaii and the University of Valladolid (UVA) in Spain.

The Principal Investigator for SuperCam is Dr. Roger Wiens of the Los Alamos National Laboratory.

Fig. 3.35 This photo shows SuperCam's mast unit prior to being installed on top of Perseverance's remote sensing mast. It fires a laser at rock or soil targets and spectroscopically analyzes the vaporized material to determine its composition. SuperCam's telescope views through a window (on the right) above a microphone (here hidden by a red protective cover) that will pick up the sounds of rocks being vaporized by the laser. The electronics are inside the gold-plated box on the left. The end of the laser peeks out from behind the left side of the electronics. Photo courtesy of CNES

IMAGE LINKS

Fig 3.1 https://images.squarespace-cdn.com/content/v1/586ec16bb3db2b558ebfec60/1595623872973-PCWE7TGCIHDAFHQVBKD5/Caltech_MagSummer2020_roverspread.jpg?format=1500w

Fig. 3.2 https://mars.nasa.gov//imgs/mars2020/rover/Mars2020_callouts_body.png

Fig. 3.3 https://mars.nasa.gov//imgs/2017/10/mars_2020_cameras_labeled_web-full2.jpg

Fig. 3.4 https://ichef.bbci.co.uk/news/976/cpsprodpb/6030/production/_113042642_24828_ksc-20200214-ph-kls01_0042_web.jpg

Fig. 3.5 https://upload.wikimedia.org/wikipedia/commons/thumb/8/80/Perseverance_first_drive_on_Mars_2021-03-04.png/1024px-Perseverance_first_drive_on_Mars_2021-03-04.png

Fig. 3.6 https://mars.nasa.gov//imgs/mars2020/rover/Mars2020_callouts_arm.png

Fig. 3.7 https://mars.nasa.gov/system/resources/detail_files/22700_PIA23311-web.jpg

Fig. 3.8 https://mars.nasa.gov/system/news_items/main_images/8556_PIA23305-16.jpg

Fig. 3.9 https://cdn.cnn.com/cnnnext/dam/assets/210224115144-04-mars-perseverance-rover-exlarge-169.jpg

Fig. 3.10 https://cdn.cnn.com/cnnnext/dam/assets/210224115141-02-mars-perseverance-rover-exlarge-169.jpg

Fig. 3.11 No image address

Fig. 3.12 https://cdn.cnn.com/cnnnext/dam/assets/210224115143-03-mars-perseverance-rover-exlarge-169.jpg

Fig. 3.13 https://static.seattletimes.com/wp-content/uploads/2021/02/Mars-2020-Spacecraft-WEB-1020x1012.jpg

Fig. 3.14 https://mars.nasa.gov/system/resources/detail_files/25158_cruise_stage.jpg

Fig. 3.15 https://mars.nasa.gov/system/resources/detail_files/25443_02_CruiseStageSeparation-1200.jpg

Fig. 3.16 https://mars.nasa.gov/system/resources/detail_files/25156_Mars_Perseverance_Trajectory_0817.jpg

Fig. 3.17 https://mars.nasa.gov//imgs/mars2020/spacecraft/backshell-web.jpg

Fig. 3.18 https://mars.nasa.gov//imgs/mars2020/spacecraft/rover-web.jpg

Fig. 3.19 https://static.scientificamerican.com/sciam/cache/file/CB9852E7-057F-4348-B686D68FCA48A5C1_source.jpg?w=590&h=800&D071D9B4-5B97-45F6-A791ACA38C3D153A

Fig. 3.20 https://www.lockheedmartin.com/content/dam/lockheed-martin/space/photo/aeroshell/Mars_2020_Heat_Shield_Mated_to_Back_Shell-NASA.jpg.pc-adaptive.1280.medium.jpg

Fig. 3.21 https://cdn.mos.cms.futurecdn.net/xNUc4eMLndwHZdffJngXad-1024-80.jpg.webp

Fig. 3.22 https://mars.nasa.gov//imgs/mars2020/powered-descent.jpg

Fig. 3.23 https://mars.nasa.gov/imgs/mars2020/mars2020-sky-crane.jpg

Fig. 3.24 https://mars.nasa.gov/system/resources/detail_files/25489_1a-EDL-Graphic_Horizontal-Imperial-01-web.jpg

Fig. 3.25 https://mars.nasa.gov/system/resources/detail_files/25045_Perseverance_Mars_Rover_Instrument_Labels-web.jpg

Fig.3.26 https://mars.nasa.gov/layout/mars2020/images/mastcam-web.jpg

Fig. 3.27 https://mars.nasa.gov/system/resources/detail_files/25283_MEDA_booms_1024.jpg

Fig. 3.28 https://mars.nasa.gov/system/resources/detail_files/25288_mars2020-2019-03-21-104942_-_D2019_0320_RL2605.jpg

Fig. 3.29 https://mars.nasa.gov/imgs/2015/07/Mars2020-MOXIE-functional-block-diagram-br.jpg

Fig. 3.30 https://mars.nasa.gov/system/resources/detail_files/25289_PIXL_2000.jpg

Fig. 3.31 https://mars.nasa.gov/system/resources/detail_files/26058_e2-PIA24095-web.jpg

Fig. 3.32 https://mars.nasa.gov/system/resources/detail_files/25281_RIMFAX_786x401.jpg

Fig. 3.33 https://mars.nasa.gov/imgs/2015/07/Mars2020-RIMFAX-antenna-prototype-br.jpg

Fig. 3.34 https://mars.nasa.gov/layout/mars2020/images/sherloc-web.jpg

Fig. 3.35 https://mars.nasa.gov/system/resources/detail_files/25290_Supercam_1280.jpg

4

Landing Site

4.1 SITE SELECTION CRITERIA

4.1.1 The Landing Site Selection Committee

Landing site selection activities got underway in earnest after the issuing of the Announcement of Opportunity for the Mars 2020 Investigation and the Science Definition Team's report in July 2013, followed by the definition of the mission initial engineering constraints and the first call for candidate landing sites. This invitation in late 2013 sought to get a jump start on the imaging of locations of high science interest using Mars Reconnaissance Orbiter (MRO) instruments to facilitate robust assessments and mature discussions of the potential merits and shortcomings for the broadest possible set of sites (relative to both engineering and science constraints) at the first workshop. At this time there was an existing database of 55 candidate landing sites.

Shortly thereafter, NASA sponsored the appointment of a Landing Site Steering Committee. Involved were Thomas Zurbuchen, Associate Administrator of the Science Mission Directorate, Michael Meyer, the Lead Scientist for the overall Mars Exploration Program, and Ken Farley, Mars 2020 Project Scientist at JPL.

In 2014, NASA appointed Dr. John A. Grant III of the Smithsonian National Air and Space Museum and Dr. Matthew P. Golombek of JPL (both of whom were geologists) as co-chairs of the Landing Site Steering Committee. The committee included additional members of the science community with a range of scientific expertise. Recognition of the need to involve additional scientists and engineers possessing experience in past and ongoing missions and site characterization and selection prompted the solicitation of, and participation by, a variety of people in the science community and at NASA. They included:

M. von Ehrenfried, *Perseverance and the Mars 2020 Mission*,
Springer Praxis Books, https://doi.org/10.1007/978-3-030-92118-7_4

- Dave Desmarais, NASA Ames Research Center.
- Brad Jolliff, Washington University St. Louis.
- Scott McLennan, SUNY Stony Brook.
- John Mustard, Brown University.
- Steve Ruff, Arizona State University.
- Kenneth Tanaka, US Geological Survey, Flagstaff.

Expertise on returned samples was incorporated via a Returned Sample Science Board (RSSB) that was appointed by NASA and the Mars 2020 Project, drawn from both Mars and non-Mars experts. In addition, the co-chairs of the Landing Site Steering Committee worked closely with NASA Headquarters, the MRO Project and the Mars 2020 Project to define the number and rate at which MRO data for candidate landing sites should be targeted and obtained. They included:

- David Beaty (Co-chair), Jet Propulsion Laboratory.
- Hap McSween (Co-chair), University of Tennessee.
- Andrew Czaja, University of Cincinnati.
- Elizabeth Hausrath, University of Nevada.
- Chris Herd, University of Alberta.
- Munir Humayun, Florida State University.
- Scott McLennan, Stony Brook University.
- Lisa Pratt, Indiana University.
- Mark Sephton, Imperial College.
- Andrew Steele, Carnegie Institute of Washington.
- Ben Weiss, Massachusetts Institute of Technology.
- Francis McCubbin (Ex officio), NASA Johnson Space Center.
- Yulia Goreva (Ex officio), Jet Propulsion Laboratory.
- Michael Meyer (Observer), NASA Headquarters.
- Betsy Pugel (Observer), NASA Headquarters.
- Lindsay Hays (Observer), Jet Propulsion Laboratory.

Over the ensuing years, hundreds of scientists and engineers supported the four Landing Site Steering Committee workshops. Also during this period the Mars Exploration Program Analysis Group (MEPAG) conducted meetings, many of which were related to the Mars 2020 mission.

4.1.2 Committee Selection Criteria

Community consensus was solicited with respect to high priority sites. Candidate sites were ranked (high, medium, low priority) based on presentations made at the workshops, then these rankings informed subsequent imaging priorities. The goal was to build up robust image datasets while orbital assets were still operating for as many high priority sites as possible, to facilitate comprehensive discussions at subsequent workshops of their relative science merits and safety issues.

Landing site prioritization is fraught with differing and deeply held opinions that are driven by personal experience, scientific taste, and varying interpretations of the existing data. As Mars 2020 was to be the first step in a possible Mars sample return effort, selecting its target had the potential to be among the most influential decisions in planetary exploration. Hence the committee co-chairs encouraged the scientists to engage in the complex and important process together and instructed them to take the long view, carefully consider all of the various stakeholders, and work with the committee to make the best assessment possible for the science of today and for future generations.

The scientists faced a few critical engineering constraints. Before looking at the science, each candidate site had to meet three basic engineering requirements. It had to be below 500 m (1,640 ft) in elevation, to provide enough atmosphere for the parachute to open. It had to be within 30 degrees of the equator to stay warm enough for the rover to survive the winter. And finally, it needed to be close to a 10 km (6.2 mi) wide flat plain which looked safe enough to land on. Once these engineering requirements were satisfied, the focus switched to which site would deliver the best science.

The criteria listed below were developed by Mars 2020 Project Science and were endorsed by the Mars 2020 Project Science Group. The criteria were designed to be as simple as possible and, following a thorough debate, scored from 1 (lowest potential) to 5 (highest potential).

Criterion 1: The site is an astrobiologically relevant ancient environment and has geological diversity with the potential to yield fundamental scientific discoveries when (a) characterized for the processes that formed and modified the geological record, and (b) subjected to astrobiologically relevant studies (e.g. assessment of habitability and biosignature preservation potential).

Criterion 2: A rigorously documented and returnable cache of rock and regolith samples assembled at the site offers the potential to yield fundamental scientific discoveries if returned to Earth in the future.

Criterion 3: There is high confidence in the assumptions, the evidence, and any interpretive models that support the assessments for Criteria 1 and 2 for the site.

Criterion 4: There is high confidence that the highest-science-value regions of interest at the site can be adequately investigated as part of the prime mission in pursuit of Criteria 1 and 2.

Criterion 5: The site has high potential for significant water resources that may be important for future exploration, whether in the form of water-rich hydrated minerals, ice/ice regolith, or subsurface ice.

In other words, the Mars 2020 mission landing site criteria asked the following questions:

- Can the Mars 2020 rover achieve all of the mission's scientific objectives at this site?
- Does the area show signs in the rock record that it once had the necessary environmental conditions to support past microbial life?
- Does the area have a variety of rocks and soils (regolith), including those from an ancient time when Mars could have supported life?
- Did geological and environmental processes, including interactions with water, alter these rocks through time?
- Are the rock types at the site able to preserve physical, chemical, mineral, or molecular signs of past life?
- Is the potential high for scientists to make fundamental discoveries with the samples that are cached by the rover, if potentially returned to Earth?
- Does the landing site have water resources (water ice and/or water-bearing minerals) that the rover could investigate in order to assess their potential use by future human explorers?
- Can the rover land and navigate without facing significant hazards posed by the terrain?

4.1.3 Mars Sample Return Considerations

Although the Mars Exploration Program includes a Mars Sample Return mission, the elements of that program have yet to be defined. The most ambitious notional MSR timeline envisions starting in 2026, with the Sample Retrieval Lander (SRL) arriving at Mars six years after the landing of the Mars 2020 rover. This schedule would allow Perseverance to make a start on the sample caching process.

The possibility of Mars 2020/SRL interaction raised the question of whether the landing site selection process should include any criteria specifically to facilitate that interaction. In the absence of concrete knowledge of the SRL capabilities, it was deemed premature to impose restrictions on landing site selection. Thus, the Mars 2020 landing site selectors should simply choose the site which offered the most scientifically compelling and diverse set of samples for potential return to Earth. In other words, the selection process should not evaluate the sites against speculative SRL capabilities.

Likewise, it was deemed premature to decide upon any specific strategy for the disposition of samples; e.g. dropping them in a depot versus retaining them on board the rover. The location of any samples that are cached on the surface will have to be coordinated between the Mars 2020 team and the Mars Exploration Program Office when the situation arises. It will surely require longer than the baseline mission of 1–1.5 Martian years for Mars 2020 to collect all 42 sample tubes. While targets remote from the landing site are of interest, in recognition that the highest probability for successful Mars 2020 mission operations occurs early in the mission, higher weight ought therefore to be given to targets which are close to the landing ellipse.

4.1.4 Landing Site Workshops

It took four Landing Site Workshops over a period of five years (2014 to 2019) for the science community to select three candidates for the Mars 2020 mission.

Workshop	Date	Results
First Landing Site Workshop	May 14–16, 2014	Initial rankings for about 30 candidate sites
Second Landing Site Workshop	Aug. 4–6, 2015	Top eight sites advanced
Third Landing Site Workshop	Feb. 8–10, 2017	Top three sites advanced
Fourth Landing Site Workshop	Oct. 16–18, 2019	Final recommendations

At the second workshop in August 2015 scientists chose eight places to study as potential landing sites for the Mars 2020 rover by asking two main questions:

- Which places would enable scientists to make the most discoveries about past life on Mars?
- Which places would give engineers the best chance of safely landing and driving the rover around?

At the third workshop in February 2017 a team of scientists narrowed down the list and chose three sites to continue as landing site candidates: Columbia Hills, Jezero Crater, and NE Syrtis.

Columbia Hills in Gusev Crater: Home to the Spirit rover

Mineral springs once burbled up from the rocks of Columbia Hills. The finding that hot springs flowed here was a major achievement by the Mars Exploration Rover, Spirit. This discovery was an especially welcome surprise because Spirit had not found signs of water anywhere else in the 160 km (100 mi) wide Gusev Crater. After the rover ceased working in 2010, studies of its older data records showed evidence that past floods may have formed a shallow lake in the crater.

NE Syrtis: Once warm, and wet

Volcanic activity once warmed NE Syrtis. Underground heat sources made hot springs flow and surface ice melt. Microbes could have flourished here in liquid water that was in contact with minerals. The layered terrain of NE Syrtis holds a rich record of the interactions which occurred between water and minerals over successive periods of early Martian history.

Jezero Crater: Wet and dry and wet again

The on-again, off-again nature of wet past of Mars is apparent from the fact that water filled and drained away from Jezero Crater on at least two occasions. More than 3.5 billion years ago river channels spilled over the crater wall and created a lake. Scientists see evidence that water carried clay minerals from the surrounding area into the crater after the lake dried up. Conceivably, microbial life could have lived here during one or more of the wet times. If so, signs of their remains might be found in lakebed sediments.

Location	Lat (°N)	Long (°E)	MOLA (km)	Ellipse Axes (km)
Columbia Hills	–14.5711	175.4374	-1.9	9 x 8
Jezero	18.4463	77.4565	-2.6	9 x 8
Midway*	18.2747	77.0480	-2.0	9 x 8
NE Syrtis	17.8899	77.1599	-2.0	9 x 8

*The Midway site between Jezero and NE Syrtis was considered as a later target of opportunity.

On Earth, the zero elevation datum is based on sea level (the geoid). Since Mars possesses no oceans it was necessary to define an arbitrary "vertical datum" for mapping the surface (areoid). In 2001, data provided by the Mars Orbiter Laser Altimeter (MOLA) was used to define the zero of elevation as the equipotential surface (i.e. gravitational plus rotational) whose average value at the equator is equal to the mean radius of the planet.

For our discussion, the important point is that all of the above sites are far below the Martian vertical datum. This is good in that landing on Mars requires a long path for all the EDL systems to work. That is one of the reasons no high altitude sites were chosen.

4.1.5 Final Recommendation

The fourth Mars 2020 Landing Site Workshop was held in Glendale, California from October 16–18, 2018 and the following is an edited summary of the debate and final recommendation in favor of Jezero Crater.

The meeting was very well attended, with 150–200 people present on each of the three days. The participants included members of the science community and the Mars 2020 project and instrument science teams on all three days. The workshop included oral presentations and discussions related to the science potential of the four remaining candidate sites Columbia Hills, Jezero Crater and NE Syrtis, plus an additional landing ellipse within the NE Syrtis region dubbed "Midway" that was closer to Jezero Crater than the original NE Syrtis ellipse and was suggested as a site by the Mars 2020 Science Team that might enable achieving the science objectives of the mission by accessing exploration targets relevant to both Jezero Crater and NE Syrtis.

Presentations emphasized the new science content, increasing confidence in the interpretations of the science potential of sites and/or potential extended mission targets for the sites. In addition, the Mars 2020 Project gave likely scenarios for each site that included discussion of potential exploration targets, observations, and sampling strategies relative to mission goals and important Mars science, as described in the 2013–2022 Planetary Science Decadal Survey (see Appendix 2 for details).

Workshop presentations were grouped into introductory sessions that related to mission status and strategies. These were followed by sessions which discussed the individual candidate sites. The final sessions on the third day were geared towards compilation of summary charts for each site and assessment relative to the science selection criteria. Additional time was allotted for discussion at the end of each session, with all discussion sessions being lively and involved. The assessment relative to the science selection criteria that followed was made using an online "ballot" submitted to Google Forms and subsequently tabulated in near real-time.

Workshop participants were instructed to assess each site relative to each of the criterion using values of 1 (lowest) to 5 (highest). An assessment tool developed by Jacob Adler at Arizona State University was used. The summary results were presented as color plots depicting the average and standard deviation of each site relative to individual criteria, and in tabular form for the average score for all six criteria for each site.

The summary rational included the following:

- The Jezero Crater site offers a river delta that extends into the crater and occurs at the same elevation as an outlet on the far side of the crater. This indicates that a lake filled the crater in Noachian to Early Hesperian times on Mars. The delta and nearby outcrops expose clays and other materials whose properties make them favorable for preserving organics and other biogenic signatures. In addition, there are carbonate-bearing rocks whose origin may relate to past weathering, and overlying cratered and possibly volcanic rocks on the crater's floor that could be used to help to constrain the chronology. Candidate extended mission options include access to the Midway site. Remaining concerns relate to the duration of the lake and its age, the origin of the carbonate-bearing rocks, and whether the crater floor rocks are of the requisite type and extent to be helpful in constraining the broader chronology of the planet.
- The Midway and NE Syrtis sites both include access to ancient rocks that show clays and carbonate-bearing rocks with evidence for past alteration by water. These rocks form a widespread and diverse sequence that spans the Noachian to Hesperian age range and may be accessible at either site. Of particular note was that both these sites expose large blocks of ancient and sometimes layered materials dubbed "megabreccia" that were likely excavated and emplaced during formation of the nearby Isidis (NE Syrtis) and/or

Jezero craters (Midway). As a result, these rocks, and others in and around the sites might provide insight into the changing conditions on the planet – ranging from those recorded in ancient subsurface groundwater environments to younger sequences possibly shaped by surface drainage. Possible extended mission scenarios include a trek to the south from the NE Syrtis site in order to study younger sulfate and volcanic rocks, to a journey from Midway to Jezero Crater to access the materials which are exposed near the delta there. However, there are multiple hypotheses for the origin of the rocks which are exposed at these sites and it is unclear how organics might be concentrated and preserved. Furthermore, high-resolution spectral data for Midway is sparse and makes understanding key aspects of the exploration targets more difficult and it is not certain whether the megabreccia blocks at Midway are in fact derived from the Isidis Basin or Jezero Crater.

The summary assessment plots revealed that the Jezero Crater site was assessed the highest, or nearly the highest, for all criteria pertaining to both the prime and extended missions. The Midway and NE Syrtis sites were also ranked highly for all criteria for the prime and extended mission, but slightly below Jezero Crater. By contrast, Columbia Hills was consistently rated lower than the other three for almost all criteria related to both the prime and extended mission objectives (the exception being it was assessed slightly higher than the Midway site in terms of confidence in its interpretation for the prime mission). Interestingly, the Midway and Jezero Crater sites were assessed the highest (and received the most votes for high potential) for the extended mission criteria, perhaps reflecting the interest in possible extended mission opportunities between the two sites.

It is apparent that the workshops brought broad expertise into assessment of the candidate landing sites and resulted in energetic discussion of the relative merits of candidate landing sites.

4.2 JEZERO CRATER

The process of landing site selection involved a combination of members of the mission team and scientists from around the world carefully studying more than 60 candidate locations on the Red Planet. After the exhaustive five-year process, NASA accepted their recommendation and named Jezero Crater as the target for the Mars 2020 Perseverance rover.

In recognition of the fact that the crater once held a lake, in 2007 it was named "Jezero" because in most Slavic languages the word means lake. It is correctly pronounced "YEH-zuh-doh" but some of the Mars 2020 mission team members say "DZEH-zuh-row" and most use a hard "J" for "JEZ-zuh-row."

Jezero Crater tells a story of the on-again, off-again nature of a wet past. More than 3.5 billion years ago river channels spilled over the crater wall and made a lake. It seems that water carried clay minerals from the surrounding area into the crater. Conceivably, microbial life could have lived in Jezero during one or more of these wet times. If so, signs of their remains might be found in either lakebed or shoreline sediments. Scientists will study how the region formed and evolved, seek signs of such life, and collect samples of rock and soil that might preserve a record of it.

Some 49 km (30 mi) wide, Jezero Crater is 2.6 km (8,530 ft) below the vertical datum on the western edge of a flat plain called Isidis Planitia (often called the Isidis Basin). To put that in perspective, Jezero could hold two cities the size of Paris (because that city is so compact) or one the size of Moscow. Only half of greater London would fit into the crater, however. Isidis is a very large impact basin approximately 1,500 km (930 mi) in diameter centered at about 87°E and 12.9°N that lies at the much lower elevation of 3.8 km (12,500 ft).

For a video by Project Scientist Ken Farley describing the reasons for selecting Jezero Crater as the Mars 2020 target, go to: https://youtu.be/qnZ_sidmr4Y

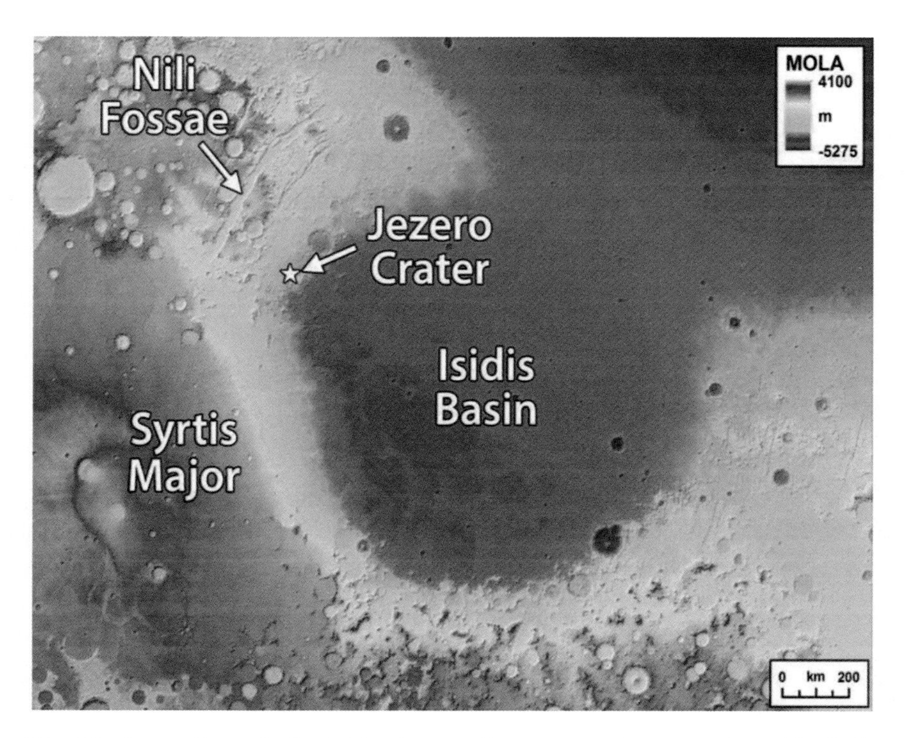

Fig. 4.1 Jezero Crater (indicated by the star) lies at the northeastern edge of the Isidis Basin. Billions of years ago, an asteroid struck Mars and excavated the crater now called the Isidis Basin. A later impact produced the much smaller Jezero Crater, with the overlapping pair of impacts uniquely changing the rocks in the region, helping to create a special landscape that scientists think may have once been friendly to life. Photo courtesy of NASA/JPL-CalTech/USGS

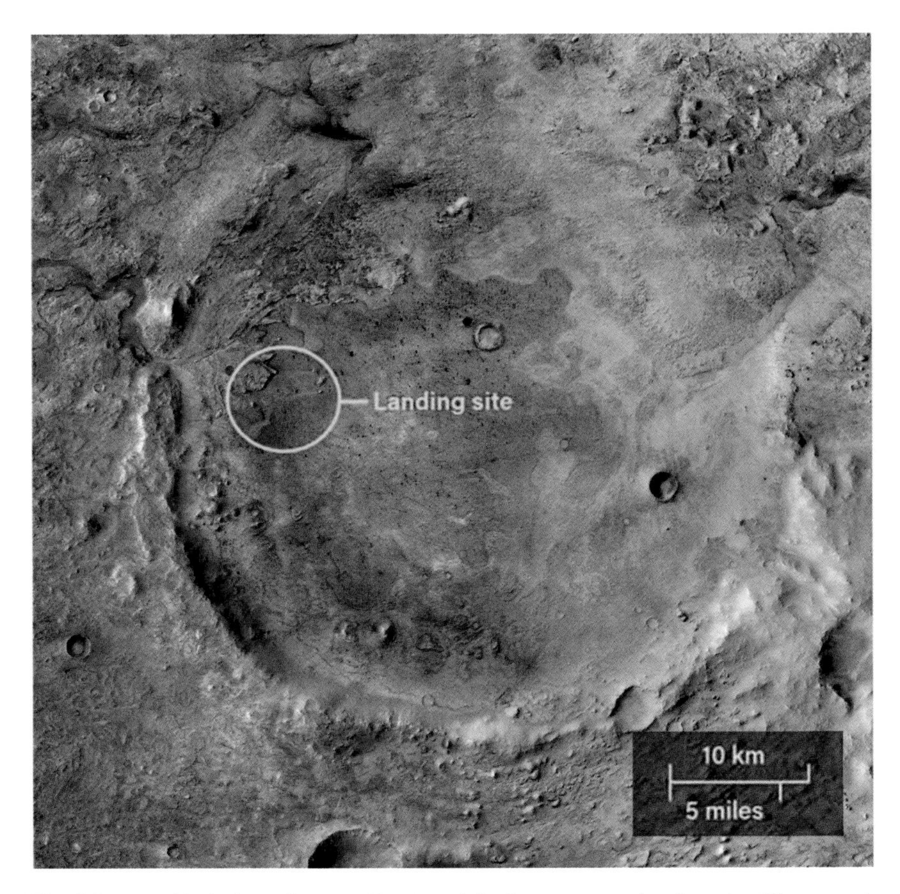

Fig. 4.2 An orbital view of Jezero Crater and the Perseverance landing site. The natural-color mosaic combines images from NASA's Mars Reconnaissance Orbiter and ESA's Mars Express. The Mars 2020 landing site (circled) is near the ancient river delta that spills over the crater's rim on the left. Scientists are eager to sample rocks at the center of the delta, where the water would have been deepest. The muddy deposits there could preserve a record of organic matter in the way that similar rocks do on Earth. Perhaps the most intriguing possibility is that Jezero Crater could have once been home to microbial mats, like the scum that forms at the edges of lakes on Earth. Certain minerals could have preserved that scum as a stromatolite, which is a kind of layered rock that is essentially a fossil. The Perseverance rover will keep a careful eye out for such structures. Photo courtesy of NASA/JPL-CalTech/MSSS/ESA/DLR/FU-Berlin/J. Cowart

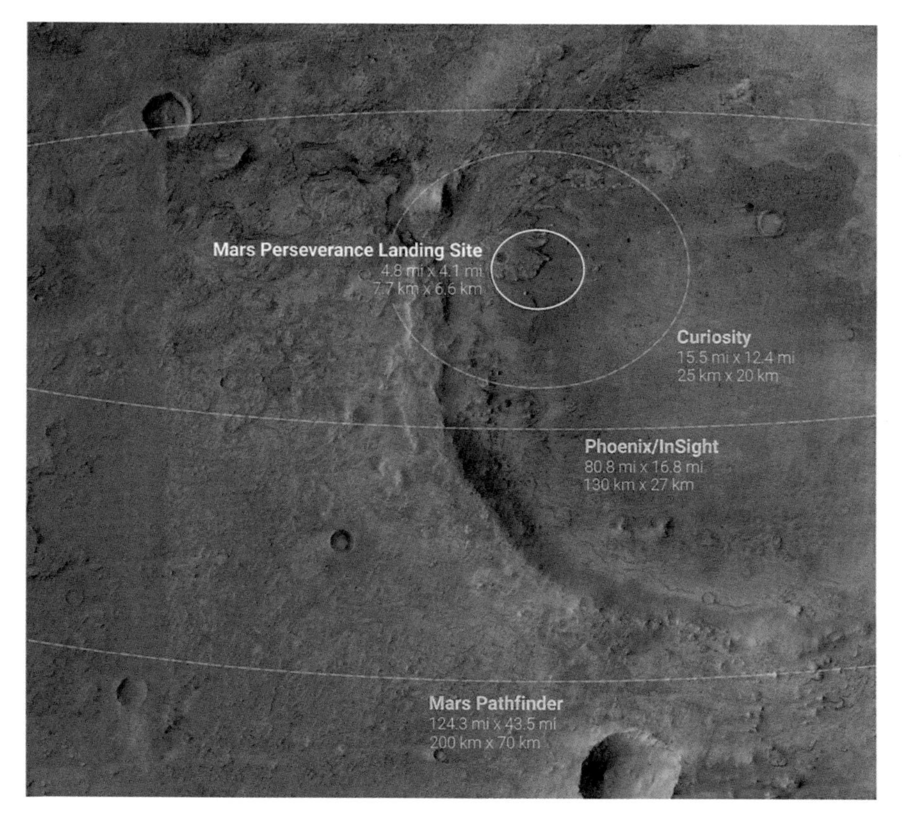

Fig. 4.3 This illustrates the improved navigation and targeting of the Mars 2020 mission compared to the much larger landing ellipses for the Curiosity, InSight, Phoenix and Mars Pathfinder missions. Photo courtesy of ESA/DLR/FU-Berlin/NASA/JPL-CalTech

Fig. 4.4 Perseverance's landing site. It was named after the science fiction author Octavia E. Butler. Butler grew up in Pasadena, California, near the Jet Propulsion Laboratory. The first African American woman to win both the Hugo and Nebula awards, and the first science fiction writer to be honored with a MacArthur Fellowship, her writing inspired many in the planetary science community and beyond. Photo courtesy of Space.com

Fig. 4.5 Perseverance's first high-resolution color image was transmitted to Earth on February 18, 2021. It was taken by the cameras on the underside of the rover. Photo courtesy of NASA/JPL-CalTech

Fig. 4.6 The Mastcam-Z imaging system captured this 360-degree panorama of "Van Zyl Overlook," where the rover was parked while the Ingenuity helicopter performed its first flights. The image was stitched together on Earth from 142 individual frames taken on Sol 3 of the mission, February 20, 2021. It reveals the crater rim and cliff face of an ancient river delta in the distance. The camera can resolve details as small as 3–5 mm (0.1–0.2 in) in the immediate vicinity of the rover and 2–3 m (6.5–10 ft) across in the distant slopes along the horizon. The detailed composite shows a Martian surface similar to images captured by previous NASA rover missions. Photo courtesy of NASA/JPL-CalTech/MSSS/ASU

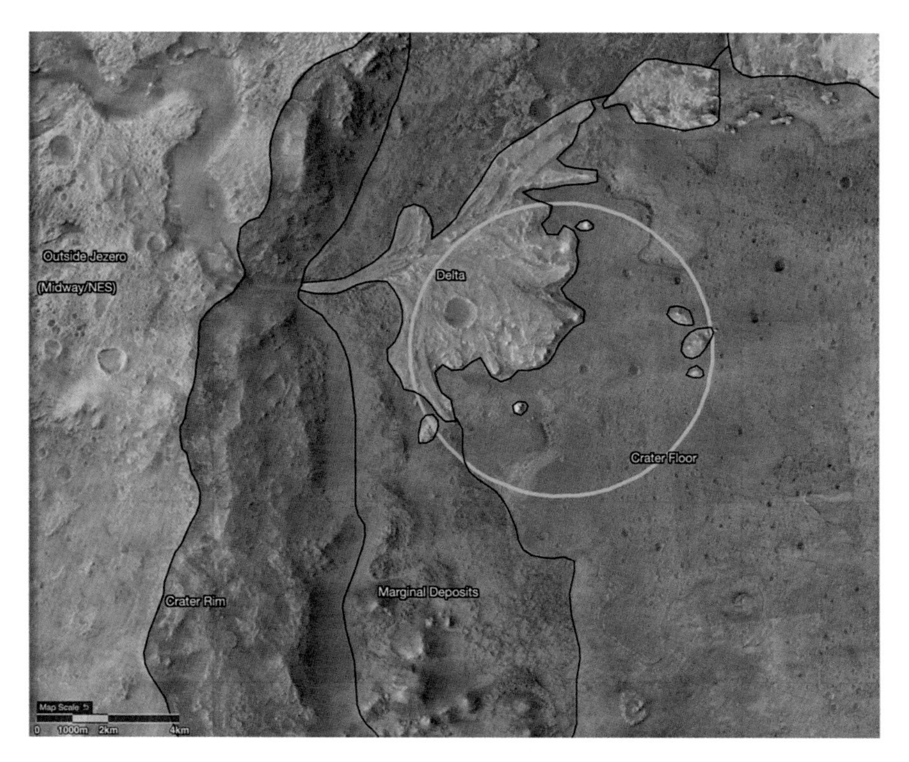

Fig. 4.7 Data from Mars Reconnaissance Orbiter was used to create this map showing regions of interest in and around Jezero Crater. The green circle represents the Mars 2020 landing ellipse. The crater held a lake and river delta billions of years ago. The Perseverance rover will seek samples of rock that may contain evidence of ancient microscopic life. Each of these regions represents a distinct area that may hold different kinds of evidence. The map was created using a tool called Campaign Analysis Mapping and Planning (CAMP) which was developed by NASA's Jet Propulsion Laboratory. Photo courtesy of NASA/JPL-CalTech/USGS/University of Arizona

4.3 SURFACE OPERATIONS

The landing location indicated using a star in Fig 4.4 marks the beginning of the Mars 2020 surface operations. Throughout, the science team will use a naming system similar to that for the locations that Curiosity is exploring at Gale Crater. Before Perseverance launched, the team mapped the entire landing site in Jezero Crater, dividing it into squares 1.2 km (0.75 mi) per side. These quadrangles were named for various national parks and preserves on Earth. As a nod to the diversity of its international science partners, the team used names from parks in countries that have contributed to the mission. As the rover explores, the team will name a succession of interesting features after inspiring locations here on Earth.

Fig. 4.8 Jezero Crater filled with water. When life was probably just gaining a foothold on Earth, Mars was young and wet and Jezero Crater held a lake some 500 m (1,600 ft) deep. Scientists think a network of rivers probably fed into the crater. The water was likely so high at one point that it spilled out over the crater walls. Scientists think it would have been an excellent place for life to evolve on the Red Planet. Photo courtesy of NASA/JPL-CalTech

After landing safely on February 18, 2021, Perseverance has a primary mission span of at least one Martian year (687 Earth days). With the science goals and objectives in mind (see Chapter 2), the JPL surface operations phase marks the time during which the rover conducts its scientific studies.

The commissioning phase began immediately after landing, with engineers and scientists spending several weeks putting the rover through its paces by testing every instrument, subsystem and subroutine, including testing the rover's drive functions. Furthermore, within 3 weeks of landing, the JPL team began the test and checkout of the Ingenuity helicopter (see Chapter 6).

4.3.1 Science Phase

On June 1, 2021, Perseverance initiated the Mars 2020 mission's science phase during which it will explore a 4 square km (1.5 square mi) patch of ground that might hold Jezero Crater's deepest and most ancient layers of exposed bedrock. This involves a number of fundamental steps:

Step 1: Find compelling rocks

As the Perseverance rover explores, scientists identify promising rock targets. Water is key to life as we know it, so they are especially looking for rocks that were formed in or were altered by water. Such rocks are even more relevant if they have organic molecules, the carbon-based chemical building blocks of life. Some special types of rocks can preserve chemical traces of life over billions of years. The key is finding rocks which formed in water and are able to preserve evidence of organics. This will greatly improve our chances of finding ancient traces of microbial life, if it ever existed. In addition to these special rocks, the rover will also collect volcanic rocks to help to establish a record of geological and environmental changes over time.

Step 2: Collect rock samples

After scientists have identified a rock target of interest, Perseverance will drill a core sample.

Step 3: Seal the rock samples

The core sample will be detached from the rock and the sample tube capped and hermetically sealed.

Step 4: Carry the samples

The rover will transport the sealed tubes in a storage rack until the mission team uses a strategy called depot caching to determine when and where to leave tubes on the Martian surface. In the baseline plan, Perseverance will place one or more large groups of samples in strategic locations. It may ultimately cache over thirty selected rock and soil (regolith) samples.

Step 5: Cache the samples

The rover will put the Martian samples, the witness blanks, and procedural blanks in the same place on the surface ready for a possible future mission to retrieve and return to Earth.

The next chapter describes the first attempt to collect such a sample, and assesses the difficulties that arose.

IMAGE LINKS

Fig. 4.1 https://www.universetoday.com/wp-content/uploads/2019/05/Jezero_crater-Isidis_basin.jpg

Fig. 4.2 https://astronomy.com/-/media/Images/News%20and%20Observing/News/2020/07/Perseverancelandingsite Jezerocrater.jpg?mw=600

Fig. 4.3 https://d2pn8kiwq2w21t.cloudfront.net/images/jpegPIA24377.width-1600.jpg

Fig. 4.4 https://cdn.mos.cms.futurecdn.net/6yVNnLQRMbzfvm5HyEo66D-1200-80.jpg

Fig. 4.5 https://www.nasa.gov/sites/default/files/styles/full_width_feature/public/thumbnails/image/pia24430-1041.jpg

Fig. 4.6 https://d2pn8kiwq2w21t.cloudfront.net/images/Vanzyl360-Ver4_Injected-web.width-1280.jpg

Fig. 4.7 https://www.al.com/resizer/wrOy02WIcQJFntHbibPQVmYQIY0=/700x0/smart/cloudfront-us-east-.images.arcpublishing.com/advancelocal/RFLX53VPPBCH5JZFPX7SWGD3SE.png

Fig. 4.8 http://www.nasa.gov/sites/default/files/thumbnails/image/pia24172.jpg

5

Surface Operations and Science

5.1 INITIAL CAMPAIGN TO THE SOUTH

On June 1, 2021 NASA's Perseverance Mars rover kicked off the science phase of its mission by leaving the "Octavia E. Butler" landing site and heading south.

Since arrival on February 18 it had undergone systems verification tests, then it had supported over a month of technology demonstration flights by the Ingenuity helicopter. The rover had tested its oxygen-generating MOXIE instrument and its cameras had returned more than 75,000 images. In addition, its microphones had recorded the first audio soundtracks of Mars.

In the first several weeks of the science campaign, the mission team planned the route complete with optional turnoffs, and labeled areas of interest and potential obstructions on the route. The path would see the rover using its auto-navigation systems to drive to a low-lying scenic overlook from which it could survey some of the oldest geological units in Jezero Crater.[1]

Perseverance explored two geological units in which Jezero's deepest (and most ancient) layers of exposed bedrock and other intriguing features are to be found. The first unit, designated "the Crater Floor Fractured Rough," is the crater-filled floor of Jezero. The adjacent unit, named "Séítah" (meaning "amidst the sand" in the Navajo language), has its fair share of the bedrock but is also home to ridges, layered rocks, and sand dunes. The goal of the initial campaign was to determine which four locations in these units best describe the story of Jezero

[1] Note that a "geological unit" is a volume of rock or ice of identifiable origin and age range that is defined by the distinctive and dominant, easily mapped and recognizable petrographic, lithologic or paleontological features that characterize it.

M. von Ehrenfried, *Perseverance and the Mars 2020 Mission*,
Springer Praxis Books, https://doi.org/10.1007/978-3-030-92118-7_5

Crater's early environment and geological history. It is believed that this area was under at least 100 m (328 ft) of water 3.8 billion years ago.

The science goals of the mission are to study the Jezero region to understand the geology and past habitability of that environment, and to search for evidence of ancient microbes. Over the course of these campaigns, the team will collect the most compelling rock and sediment samples, and the rover will cache them for possible future shipment to Earth for detailed investigations. Perseverance will also take measurements and test technologies designed to support future human and robotic exploration of Mars.

Most of the challenges along the way will be the sand dunes within the mitten-shaped Séítah unit. To negotiate them, the rover team decided to drive mostly either on the Crater Floor Fractured Rough or along the contact line between it and Séítah. When there is occasion to do so, the rover will perform a "toe dip" into the Séítah unit.

The first science campaign will conclude when the vehicle returns to its landing site. By that time, it will have traveled between 2.5–5 km (1.6–3.1 mi). Although the team initially thought that they might take as many as eight samples of rock and soil during the first campaign, the first attempt to get a sample on August 6, 2021 was a failure (see Sect. 5.1.3).

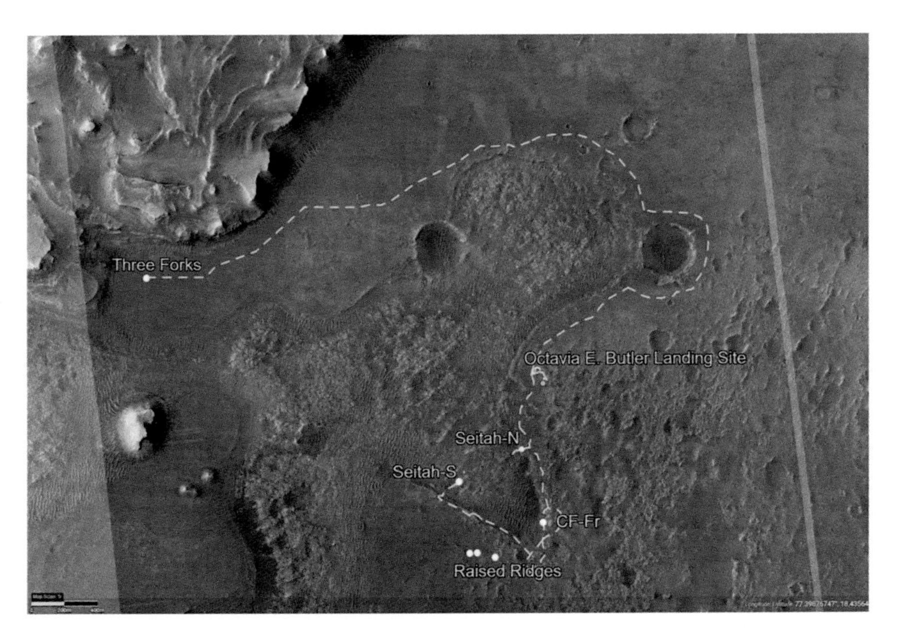

Fig. 5.1 This annotated Mars Reconnaissance Orbiter image of Jezero Crater depicts the route for Perseverance's first science campaign, which began with driving southward about 1 km (0.62 mi) onto the "Crater Floor Fractured Rough" (CF-Fr) geological unit. The map also shows a possible route to the north and west of the landing site for the second science campaign. (Photo courtesy of NASA/JPL-CalTech/University of Arizona)

Fig. 5.2 Perseverance looks back toward its tracks on July 1, 2021 (sol 130) after driving autonomously for 109 m (358 ft); its longest such drive to date. Taken by one of the NavCams, the image has been processed to enhance the contrast. (Photo courtesy of NASA/JPL-CalTech)

On August 6, Perseverance reached what its operating team called "paver stones" which were flat, white, dust-coated rocks found throughout much of the floor of Jezero Crater. This terrain is believed to be the oldest in the crater. But it wasn't yet apparent whether this landscape was deposited by the lake or was formed by volcanic flows. Close-up images failed to resolve whether the paver stones were of igneous or sedimentary origin. They are covered with sand grains and pebbles, along with some sort of purplish coating that impaired remote measurements. It was at this location that the rover made its first attempt to acquire a core sample.

After discussing whether additional exploration farther south and then west was warranted, the team decided that Perseverance should retrace its steps. Its return to the Octavia B. Butler landing site would conclude the first science campaign.

Fig. 5.3 The area in which Perseverance was to seek its first sampling target. The 28 individual images that were combined to make the larger main image were taken by the Mastcam-Z on July 8, 2021 (sol 136). The image calibration depicts "natural color" to simulate the approximate view we would see with our own eyes. (Photo courtesy of NASA/JPL-CalTech/ASU/MSSS)

Fig. 5.4 Paver Stones. (Photo courtesy of NASA/JPL-CALTECH/MSSS)

Several months of travel lay ahead with Perseverance making its way to "Three Forks," where the second science campaign was to begin. From that position it will have access geology at the base of the ancient delta (the fan-shaped relic of the confluence of a river with a lake) as well as being able to ascend the delta by driving up a valley to the northwest.

5.1.1 Sample Caching

The Sample Caching System (SCS) of the Mars 2020 mission is best understood by watching NASA/JPL engineers testing it, as in this 2:34 minute video:

https://mars.nasa.gov/system/video_items/5953_JPL-20200529-MARSf-0001-Mars_Sample_Caching-MEATBALL-1080.mp4

The following is an edited transcript of the video:

Chief Engineer, Adam Steltzner: "In terms of robots that go into space, the SCS on the Mars 2020 mission is the most complicated, most sophisticated thing that we know how to build. This is a system which allows us to take core samples of rocky material on the surface of Mars, carefully seal them in very sterile vessels for eventual return to Earth. We've been working on this system for seven years, and that is because it's a tough job."

Integration & Test Lead, Kelly Palm: "We're testing the equipment to make sure that it is going to work, when we get to Mars. It has to function on its own. We have to think of all eventualities, and try them here first. And then, if they don't work, change it now, because we can't make any changes later."

Adam Steltzner: "To drill into the rock on Mars, pull out intact core samples and seal them hermetically, and for this to be done autonomously by a robot hanging off the end of a rover on the surface of Mars, has been a challenge. We have got actually three robots necessary to do the sample and caching. The big robotic arm out on the front of the rover takes our drill and pushes it against the surface. That allows us to take core samples. Then we put that core sample in the carousel. The second robot takes it from the robotic arm and puts it inside our adaptive caching system. This is the part of the sample and caching system inside the rover. We've got a little tiny robot, a special robot arm called the SHA – the Sample Handling Arm. It takes the samples out of the carousel, and moves them through volume-assessment, image-taking, and eventually to sealing. It then replaces the cylinder containing the sample in a storage spot. It does this all on its own in the matter of a few hours. We have designs for bringing them back in a decade. Mars has been at the fore of our consciousness about the questions of life. Could life exist at one of our nearest neighbors? I think we have a lot to learn (life or no life) about the evolution of our solar system, and about our planet, by looking in-depth at rocks brought back from Mars."

5.1.2 Sampling Approach

Given the importance of acquiring samples that can eventually be flown to Earth for analysis, the Perseverance rover will explore many sites to obtain samples of different types of terrain. This mix of samples may determine the nature of both igneous and sedimentary deposits.

Neil Armstrong took only 3 minutes 35 seconds to collect the first sample on the Moon, a scoop of regolith that was taken as a "contingency" against any problem which meant he had to retreat to the Lunar Module. Perseverance required about 11 days to complete its first sampling attempt because its Sampling and Caching System is the most complex and capable mechanism ever sent into space.

Fig. 5.5 This illustration shows Perseverance ready to sample. The sequence begins with the rover placing everything necessary for sampling within reach of its 2 m (7 ft) long robotic arm. It then performs an imaging survey to enable the science team to determine the precise location for taking the sample and select a separate target site in the same area for "proximity science." (Illustration by NASA/JPL)

"The idea is to get valuable data on the rock we are about to sample by finding its geological twin and performing detailed in-situ analysis," said science campaign co-lead Vivian Sun. "On the double, first we use an abrading bit to scrape off the top layers of rock and dust to expose fresh, unweathered surfaces, blow it clean with our Gas Dust Removal Tool, and then get 'up close and personal' with our turret-mounted proximity science instruments SHERLOC, PIXL and WATSON" whose cameras will provide mineral and chemical analysis of the abraded target. "SuperCam and Mastcam-Z, both located on the rover's mast will be used. While SuperCam fires a laser at the abraded surface and spectroscopically measures the resulting plume and collects other data, Mastcam-Z will capture high-resolution imagery. Working together, these five instruments enable unprecedented analysis of geological materials at the worksite. After the pre-coring science is complete, we will allow the rover to fully charge its battery for the events of the following day."

Sampling day kicked off with the sample-handling arm in the Adaptive Caching Assembly (ACA) retrieving a sample tube, heating it, and then inserting it into a coring bit. A device called the bit carousel transported the tube and bit to a rotary-percussive drill on Perseverance's robotic arm, which then drilled the untouched geological "twin" of the rock studied the previous sol, filling the tube with a core sample roughly the size of a piece of chalk. The arm then moved the bit-and-tube combination back into the carousel, which transferred it back into the ACA, for the sample to be measured for volume, photographed, sealed, and stored.

"Not every sample Perseverance is collecting will be done in the quest for ancient life, and we don't expect this first sample to provide definitive proof one way or the other," said Project Scientist Ken Farley of CalTech. "While the rocks located in this geological unit are not great time capsules for organics, we believe they've been around since the formation of Jezero Crater and thus are incredibly valuable for filling gaps in our geological understanding of this region, things that we will desperately need to know if we find life once existed on Mars."

5.1.3 First Sampling Attempt at Cratered Floor Fractured Rough

After the commands for the first sample acquisition and processing on the target "Roubion" were sent the team took several hours off to wait for the result. More than 90 engineers and scientists who had worked years in preparing for this moment congregated online at 2 a.m. PDT on Friday, August 6, 2021 to receive the news of the coring operation. It verified that the corer had achieved the full commanded depth 7 cm (2.76 in) and imagery showed the hole in the rock to be surrounded by the tailings produced around the borehole during coring.

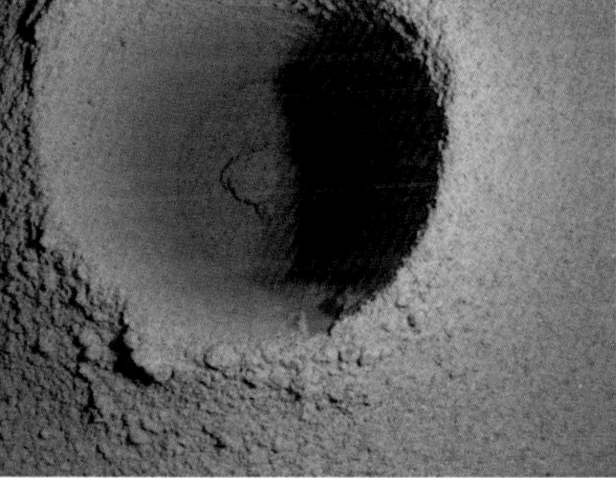

Fig. 5.6 These images of "Roubion" showing the drill hole for the first sample-collection attempt were taken on August 6, 2021. The left image with the shadow of the rover was taken by a NavCam. The right image is a composite generated from multiple images taken by the WATSON camera. (Photo courtesy of NASA/JPL-CalTech/MSSS)

The next morning was a rollercoaster of emotions. Engineering telemetry and an image from the CacheCam in the tube processing hardware confirmed the system had transferred the sample tube from the corer to the ACA, sealed it and placed it in storage. This was a magnificent first-time success! The team was elated. Then the volume measurement and post-measurement image arrived indicating that the sample tube was empty. It took several minutes for this reality to sink in, then the team transitioned to investigation mode.

Fig. 5.7 This composite of four images of "Roubion" was taken by SuperCam from a distance of 2.23 m (7.32 ft) on August 8, 2021. The small pits within the hole were made by laser zaps from SuperCam during later efforts to analyze the rock's composition. The science team concluded that owing to the rock's unusual composition, the process of extracting a core created a significant pile of tailings around the coring hole. Eight pits produced by 30 laser shots each are seen in two columns inside the drill hole. The SuperCam team's analysis suggests that the top six pits penetrated the compacted mound of tailings, while the bottom two pits in the hole interrogated material below the rock surface. Two additional laser pits can be seen in the tailings at the near side of the hole. Two vertical ridges inside the hole, one on each side of the laser pits, were produced as the drill was being removed, prior to laser analysis. Some bright mineral grains can be seen as glints in the tailings and in the hole. A few clumps or larger pieces of material are seen at the top of the tailings pile just to the left of the hole. (Image courtesy of NASA/JPL-CalTech/LANL/CNES/IRAP)

5.1.4 Assessment

What followed was two full days of combing through the data about the first sampling attempt and adding more observations to the tactical plan to aid the investigation.

According to Louise Jandura, Chief Engineer for the Sampling and Caching System:

- Engineering telemetry of the Corer performance during both the abrasion and coring activities did not uncover any unusual responses compared to the data from our successful Earth-based test of more than 100 cores in a variety of test rocks.
- Imaging of the workspace in the areas over which the hardware traveled during the post-coring activities did not result in finding an intact core or core pieces on the Martian surface.
- Depth measurements of the borehole derived from the merging of image products from WATSON, along with the imagery itself, led the team to believe that the coring activity in this unusual rock had resulted only in powder and small fragments which were not retained owing to their size and the lack of any significant chunk of a core.
- It appears that the rock was simply not robust enough to produce a core. Some material is visible in the bottom of the hole. The material from the desired core is likely either in the bottom of the hole, in the tailings that surround the hole, or some combination of both.
- Both the science and engineering teams concluded that the uniqueness of this rock and its material properties were the dominant contributor to the failure to extract a core from it.

Despite this disappointing result, the team celebrated the first fully autonomous sequence of sampling system on Mars within the time constraint of a single sol.

And in any case, the tube holds a sample of the Martian atmosphere, a task that had originally been assigned to later in the mission.

5.1.5 Second Attempt in South Séítah

Around the middle of August 2021 Perseverance left the Cratered Floor Fractured Rough terrain and headed towards "South Séítah," the farthest point of this phase of the mission, where they expected to find sedimentary rocks more like those that were drilled during pre-mission tests.

As of August 19, Project Manager Jennifer Trosper pointed out that they "were currently focusing on the appropriate next steps to achieve a core sample as soon as is prudent while managing the risk of how we do it. Our first step is somewhat obvious; find a more resistant rock that is less likely to crumble for coring. There is extensive outcropping of this kind of rock all along our planned traverse route to South Séítah. This outcrop was previously identified as a high science-value target for sampling." She went on, "The plan is to select a suitable rock near the region named Citadelle. We'll first abrade the selected rock and use the science instruments to confirm that the new target is likely to result in a core sample. If we choose to sample the rock, Perseverance will perform a set of activities very close to what was done on the earlier coring target. The main difference will be, after coring, we've added a humans-in-the-loop session to review the images of the tube in the bit in order to confirm that a sample was collected. Then the tube will be transferred into the rover for processing."

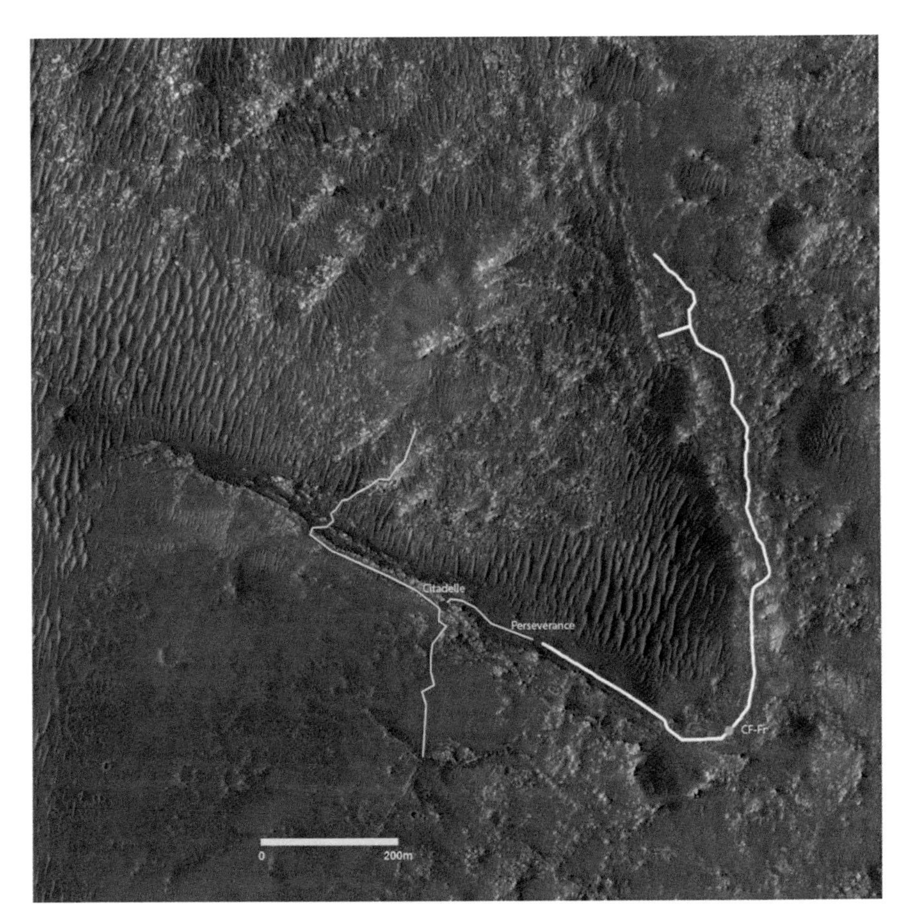

Fig. 5.8 This Mars Reconnaissance Orbiter image shows the track (indicated in white) of the Perseverance rover since it arrived on Mars on February 18, 2021. It made its first attempt at collecting a sample in the "Crater Floor Fractured Rough" area (labeled CF-Fr). The location named "Citadelle" lies on the projected track (yellow). (Photo courtesy of NASA/JPL-CalTech/University of Arizona)

Fig. 5.9 Ingenuity scouting for Perseverance. This image of sand dunes, boulders and rocky outcrops of "South Séítah" was captured by the Ingenuity helicopter from an altitude of 10 m (33 ft) on August 16, 2021, during its 12th flight. The image was one of ten collected during the flight at the request of the Perseverance science team in order to help determine whether to explore the location further. Ingenuity's shadow is visible in the lower part of the image. (Photo courtesy of NASA/JPL-CalTech)

From the attempted sample site on the Crafter Fractured Floor Rough geological unit Perseverance drove 455 m (1,493 ft) along a ridge named Artuby to the area named "Citadelle." Being the French word for castle, this was a reference to how this craggy spot overlooks Jezero Crater's floor. The ridge is capped with a layer of rock that had evidently resisted wind erosion.

Following the additional step in the sampling process of having the Mastcam-Z camera peer inside the sample tube to verify that it was not empty, the tube was sealed and stored.

Orbital imagery showed that the boulders at Citadelle were part of an extensive outcrop on the summit and far side of the ridge. They provided good targets for another coring attempt because they are very solid in appearance; a conclusion supported by the fact that even after eons of erosive wind action they still stand high in the landscape. They also had high scientific value as a potential crater-floor sample suite.

Fig. 5.10 Perseverance checks out the rock named "Rochette" in the "Citadelle" area. The image was taken on about sol 187 by one of the HazCams. The site is part of a 400 m (1,300 ft) long ridgeline called Artuby that contains rock outcrops and 10 m (33 ft) scale boulders, some of which had intriguing layering. (Photo courtesy of NASA/JPL-CalTech)

Upon arrival at Citadelle, the science operations team developed a new sampling procedure, referred to as the "sampling sol path." This outlines the standardized sequence of activities to be undertaken for each sampling event, a process which takes over a week to complete.

The sampling sol path includes the following key milestones:

- Drive up to the intended sampling location.
- Perform reconnaissance observations using the science instruments and cameras.
- Create an abrasion patch by employing the drill to grind away the upper centimeter of the rocky surface of an interesting rock.
- Analyze the fresh rock surface inside the abrasion patch with the science instruments.
- Select a nearby rock target for the coring operation.
- Use the drill to extract a core sample.
- Deliver the core (in its tube) to the belly of the rover, where it passes through a complex sequence of sample assessment, imaging, hermetic sealing and eventual storage.
- Wrap up the science observations at the sampling location, including analysis of the newly drilled borehole.
- Drive away from the sampling site with Perseverance one core sample heavier.

The sampling sol path is all about efficiency. Many activities are crammed into the sol path and the goal is to organize it into a logical sequence that maximizes rover resources. It also allows the team to acquire a standard and comparable set of scientific observations, in order to consistently document each sample that is collected.

Fig. 5.11 The abrasion patch on the rock named "Rochette." (Photo courtesy of NASA/JPL-CalTech)

Fig. 5.12 Obtained on September 1, 2021 (sol 190) this is a composite of two images of "Rochette" showing the hole named "Montdenier" that Perseverance drilled during its second sample-collection attempt. Notice the drill's dust has partially filled the abrasion site. (Photo courtesy of NASA/JPL-CalTech)

Fig. 5.13 This Mastcam-Z image shows the sample of rock from "Montdenier" inside the sample tube. The image was taken after coring concluded but prior to an operation that vibrates the drill bit and tube in order to clear the tube's lip of any residual material. The bronze-colored outer ring is the coring bit. The lighter-colored inner ring is the open end of the sample tube. The tube contains a rock core sample slightly thicker than a pencil. (Photo courtesy of NASA/JPL-CalTech/ASU/MSSS)

Having taken the requisite images, Perseverance began a procedure known as "percuss to ingest," which vibrates the drill bit and tube for one second on five separate times in order to clear residual material from the lip of the sample tube. The action can also cause a sample to slide down farther into the tube. After the percuss-to-ingest procedure was completed, a second set of Mastcam-Z images were taken, but the lighting was poor and internal portions of the tube were not visible.

"The project has got its first cored rock under its belt, and that is a phenomenal accomplishment," said a delighted Project Manager Jennifer Trosper. "The team determined a location and selected and cored a viable and scientifically valuable rock. We did what we came to do!"

"Getting the first sample under our belt is a huge milestone," said Perseverance Project Scientist Ken Farley. "When we get these samples back on Earth they're going to tell us a great deal about some of the earliest chapters in the evolution of Mars. But however geologically intriguing the contents of

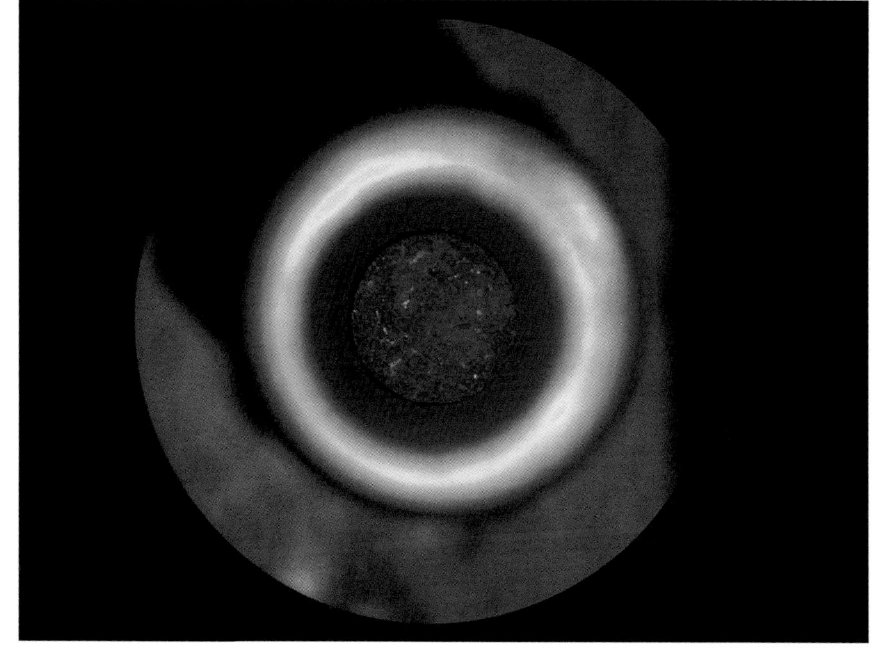

Fig. 5.14 First successful sample. The cored sample of Mars rock is visible (at center) inside a titanium collection tube in this image from the Sampling and Caching System Camera (known as the CacheCam) of the Perseverance rover. The image was taken on September 6, 2021 (sol 194), prior to the system attaching and sealing a metal cap onto the tube. It was taken so that the cored-rock sample would be in focus. The seemingly dark ring surrounding the sample is a portion of the tube's inner wall. The bright gold-colored ring surrounding the tube and sample is an asymmetrical flange called the "bearing race" which assists in shearing off a sample once the coring drill has bored into a rock. The outermost, mottled-brown disk in this image is a portion of the sample handling arm inside the Adaptive Caching Assembly. (Photo courtesy of NASA/JPL-CalTech)

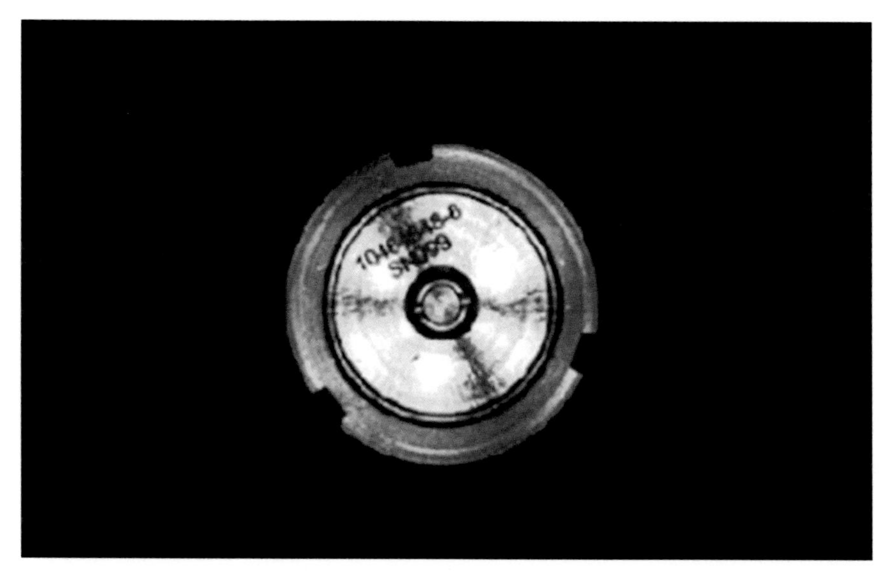

Fig. 5.15 First sample sealed. Perseverance's first cored sample of Mars rock is hermetically sealed inside its titanium tube. The item and serial numbers can be seen near the center of the disk. (Photo courtesy of NASA/JPL-CalTech)

sample tube 266 will be, they won't tell the complete story of this place. There's a lot of Jezero Crater left to explore, and we will continue our journey in the months and years ahead."

Fig. 5.16 The second drill hole in "Rochette." (Photo courtesy of NASA/JPL-CalTech)

Perseverance took a second sample, named "Montagnac," from the same rock on September 8, 2021.

5.1.6 Assessment

"It looks like our first rocks reveal a potentially habitable sustained environment," said Project Scientist Ken Farley after reviewing the evidence from Jezero Crater obtained so far. "It's a big deal that the water was there a long time."

For a 1 hr. 6 min video of NASA's conference on September 10, 2021 about the drilling and sampling the rock named Rochette, go to:

https://youtu.be/IMyuOBexwE0

The rock from which Perseverance's first core samples were obtained is basaltic in composition and may be the product of lava flows. The presence of crystalline minerals in volcanic rocks is very helpful in radiometric dating, because it helps accurately date when the rock formed. Each sample can serve as part of a larger chronological puzzle. Put them in the right order and you have a timeline for the formation of Jezero Crater, for the emergence and disappearance of its lake, and for how the Martian climate changed in the ancient past.

What is more, salts have been detected that may have formed when groundwater flowed through and altered the original minerals in the rock, or more likely when water evaporated and left behind the salts. The salt minerals in the first two rock cores may also contain trapped tiny bubbles of ancient Martian water. If present, these could serve as microscopic time capsules, offering clues about the ancient climate and habitability. And of course, terrestrial salts are well-known for their ability to preserve signs of ancient life.

The Perseverance team were aware that Jezero Crater was once filled by a lake, but for how long this existed was uncertain. The scientists could not dismiss the possibility that the lake was a "flash in the pan" of floodwater that rapidly filled the impact crater and dried up after only a few years. But the level of alteration present in the rock from which the core samples were obtained, as well as in the rock the team targeted on their first (failed) sample-acquisition attempt, implies that groundwater was present for a long time. This groundwater could have been directly related to the lake, or it could have traveled through the rocks long after the lake had dried up. Though scientists still cannot say whether any of the water that altered these rocks was present for tens of thousands or for millions of years, it is looking increasingly likely it was long enough to make the area attractive to microscopic life in ancient times.

"These samples have high value for future laboratory analysis on Earth," said Mitch Schulte of NASA Headquarters, the mission's Program Scientist. "One day, we may be able to work out the sequence and timing of the environmental conditions that the minerals in this rock represent. This will help us answer the 'big picture' science question about the history and stability of liquid water on Mars."

5.1.7 Third Attempt

Having finished sampling at the Citadelle, Perseverance headed toward the next scientific area of interest in South Séítah. It celebrated its 200th sol on Mars on September 11, 2021 with a record breaking 175 m (574 ft) drive northwest along Artuby ridge, a series of layered outcrops which outline the southern edge of the Séítah "thumb" and possibly represent a boundary between two geological units. Perseverance took the wheel for most of the drive, covering 167 m (548 ft) with its advanced AutoNav, a mobility software that maps terrain and avoids hazards for longer drives.

When Perseverance arrived at the outcrop on sol 204 on September 15, the rock that caught the science team's eye was a thinly layered outcrop that they named "Bastide." The thin layering suggested it might be sedimentary and deposited in Jezero Crater's lake over 3 billion years ago, but further data would be needed to confirm its origin. The team abraded the rock to reveal a fresh surface, to better investigate the composition using the sophisticated suite of science instruments. But with solar conjunction set to start in early October, it was time to "park the car" for an intermission.

5.1.8 Conjunction

Activities were halted during the solar conjunction moratorium on commanding all Mars spacecraft between October 2 and 14, 2021 (plus or minus a couple of days) while Mars was within 2 degrees of the Sun from our perspective. During this time an eruption of solar plasma can corrupt communications and potentially endanger a mission, so the Perseverance and Curiosity rovers and InSight lander recorded data pending the restoration of reliable communications. This passively acquired data included:

- Perseverance was to take weather measurements using its MEDA (Mars Environmental Dynamics Analyzer) sensors, look for dust devils with its cameras (though it wouldn't move its mast), operate its RIMFAX (Radar Imager for Mars' Subsurface Experiment) radar, and capture new sounds with its microphones.
- The Ingenuity helicopter would remain in position 175 m (575 ft) away from Perseverance and communicate its status to the rover on a weekly basis.
- Curiosity was to obtain weather measurements using its REMS (Rover Environmental Monitoring Station) sensors, monitor radiation with its RAD (Radiation Assessment Detector) and DAN (Dynamic Albedo of Neutrons) sensors, and use its suite of cameras to look for dust devils.
- The stationary InSight lander would continue to use its seismometer to detect Marsquake temblors.
- In addition to gathering their own science data, the Mars Odyssey, Mars Reconnaissance Orbiter and MAVEN orbiters would continue relaying a certain amount of data from the surface missions to Earth.

Although some science data would reach Earth during conjunction, the missions would store most of it until after the moratorium, at which time the Deep Space Network would spend about a week downloading recorded data prior to picking up normal spacecraft operations.

5.2 SECOND CAMPAIGN TO THE DELTA REGION

Once Perseverance has finished sampling in the South Séítah area, it is to return to the Octavia E. Butler landing site, probably by retracing its tracks. After that, the plan is to drive north and then west to reach the region of the second science campaign, the "Three Forks" area in the mouth of the river that created the fan-shaped delta. This may be especially rich in carbonates, minerals which on Earth are associated with biological processes and are able to preserve signs of ancient life. This second science campaign should be underway in early 2022.

Fig. 5.17 A possible route for the second science campaign. The green line starts in the "Three Forks" area, goes west up onto the delta, then up the valley of the Neretva river. Alternatively the science team may decide to take a route from the delta to the south. (Photo courtesy of NASA/JPL-CalTech/University of Arizona)

Perseverance may well live as long or perhaps longer than Curiosity and will be constrained by the number of sample tubes. After it has filled the tubes, the task will be to determine where to cache them for the possible retrieval and return to Earth. After it has dropped its cache, Perseverance will likely keep investigating using its onboard science instruments.

IMAGE LINKS

Fig. 5.1 https://mars.nasa.gov/system/resources/detail_files/25965_e2-PIA24596_web.jpg

Fig. 5.2 https://mars.nasa.gov/system/resources/detail_files/26050_PIA24741-2560x1920.jpg

Fig. 5.3 https://d2pn8kiwq2w21t.cloudfront.net/images/jpegPIA24746.width-1600.jpg

Fig. 5.4 https://www.sciencemag.org/sites/default/files/styles/article_main_image_-_1280w__no_aspect_/public/PIA24748-1280x720.jpg?itok=8_Pg4Z6q

Fig. 5.5 https://static.scientificamerican.com/sciam/assets/Image/2021/PIA23491_1_MSR_a_Mars_2020_collecting_sample.png?w=&fit=bounds

Fig. 5.6 https://encrypted-tbn0.gstatic.com/images?q=tbn:ANd9GcTJf_NFKPpdjq18hI99i3K6WOhhPPUHoHHM9A&usqp=CAU

Fig. 5.7 https://mars.nasa.gov/system/resources/detail_files/26132_PIA24749_web.jpg

Fig. 5.8 https://mars.nasa.gov/system/resources/detail_files/26147_PIA24800-web.jpg

Fig. 5.9 https://mars.nasa.gov/system/resources/detail_files/26177_PIA24801-web.jpg

Fig. 5.10 https://mars.nasa.gov/system/resources/detail_files/26193_PIA24767-web.jpg

Fig.5.11 https://mars.nasa.gov/mars2020-raw-images/pub/ods/surface/sol/00185/ids/edr/browse/ncam/NRF_0185_0683366539_317ECM_N0070000NCAM00303_07_195J01_1200.jpg

Fig. 5.12 https://mars.nasa.gov/system/resources/detail_files/26209_PIA24805-web.jpg

Fig. 5.13 https://mars.nasa.gov/system/resources/detail_files/26210_PIA24804-web.jpg

Fig. 5.14 https://mars.nasa.gov/system/news_items/main_images/9029_PIA24806_web.jpg

Fig. 5.15 https://mars.nasa.gov/system/resources/detail_files/26220_cachecam_web.jpg

Fig. 5.16 https://mars.nasa.gov/system/news_items/main_images/9036_1-PIA24840-web.jpg

Fig. 5.17 https://static.scientificamerican.com/sciam/assets/Image/2021/e1-PIA24379-Ken_Farley_traverse_copy.jpg?w=&fit=bounds

6

Ingenuity

6.1 OUTLINE

6.1.1 Ingenuity's Sponsors and Builders

The Ingenuity Mars Helicopter was developed by JPL, which also carried out the flight operations demonstration activities. Ingenuity was supported by two NASA Headquarters organizations: the Science Missions Directorate and the Aeronautics Research Missions Directorate. The Ames Research Center and Langley Research Center provided significant flight performance analysis and technical assistance. AeroVironment Inc., Qualcomm and SolAero also provided design assistance and major vehicle elements. Lockheed Martin Space designed and supplied the Mars Helicopter Delivery System. David Lavery was the Program Executive at NASA Headquarters. Appendix 3 includes brief résumés of many of the people involved both from Headquarters and JPL.

6.1.2 A Brief History

The first powered controlled flight on Earth occurred on December 17, 1903 on the windswept dunes of Kill Devil Hill near the township of Kitty Hawk, North Carolina. Built by Orville and Wilbur Wright, the "Flyer" covered a distance of 120 feet in 12 seconds with Wilbur having won a coin toss for who would have the honor of achieving this historic flight.

A small section of the material which covered one of the wings of the Wrights' aircraft is now onboard Ingenuity. An insulated tape was used to wrap the tiny swatch of fabric around a cable located underneath the helicopter's solar panel. The Wrights had started to use this unbleached muslin, known as "Pride of the

M. von Ehrenfried, *Perseverance and the Mars 2020 Mission*,
Springer Praxis Books, https://doi.org/10.1007/978-3-030-92118-7_6

West" to cover the wings of gliders in 1901. The Apollo 11 crew took another piece of the material, along with a small splinter of wood from the Flyer to the Moon and back in July 1969.

6.1.3 Conceptual Design

Ingenuity's design was driven by the conditions in which it had to fly. The lower gravity of Mars (about one-third of Earth's) only partially offsets the thinness of the 95% carbon dioxide atmosphere of Mars, which makes it much harder for an aircraft to generate lift. The atmospheric density of Mars is about 1/100th that of Earth at sea level or approximately the same as at 27,000 m (87,000 ft), which is an altitude that cannot be approached by even the most powerful helicopters. To keep Ingenuity aloft, its specially shaped blades of enlarged size must rotate at a speed of at least 2,400 and up to 2,900 rpm; 10 times faster than what is required on Earth. It has contra-rotating coaxial rotors some 1.2 m (4 ft) in diameter, with each rotor controlled by a separate swashplate that can affect both collective and cyclic pitch.

Ingenuity was designed to be a technology demonstrator by JPL to assess whether such a vehicle could fly safely. Before it was built, launched and landed, scientists and managers expressed a hope that helicopters could provide improved mapping and guidance in order to give future mission controllers more information to help with travel routes, planning, and hazard avoidance. Based on the performance of previous rovers through Curiosity, it was assumed that such aerial scouting might enable future rovers to drive up to three times as far per sol, but the AutoNav of Perseverance significantly reduced this advantage by allowing the rover to cover more than 100 m per sol.

The history of the Mars Helicopter team dates back to 2012, when JPL engineer MiMi Aung was leading Dr. Charles Elachi, at that time the director, on a tour of the Autonomous Systems Division. Looking at the drones demonstrating onboard navigation algorithms, Elachi asked Aung, "Hey, why don't we do that on Mars?" Engineer Bob Balaram briefed Elachi on the feasibility, and a week later Elachi told him, "Okay, I've got some study money for you." By January 2015, NASA agreed to fund the development of a full-size model which became known as the "risk reduction" vehicle.

NASA/JPL and contractor AeroVironment, Inc., famous for its unmanned aerial vehicles, published the conceptual design in 2014 for a "scout helicopter" to be deployed by a rover. By mid-2016, $15 million was being requested to continue development. By the end of 2017 engineering models of such a vehicle had been tested in a simulated Martian atmosphere and models were undergoing testing in the Arctic, but its inclusion in the Mars 2020 mission had not yet been approved.

The US federal budget, announced in March 2018, provided $23 million for the helicopter for one year and it was announced on May 11 that a helicopter was to be developed and tested in time for the mission. After extensive flight dynamics and environment testing, the finished helicopter was mounted on the belly of the Perseverance rover in August 2019. NASA spent about $80 million to create the vehicle and about $5 million to operate it.

After a public contest, in April 2020 NASA accepted the suggestion of Vaneeza Rupani, a girl in the 11th grade at Tuscaloosa County High School in Northport, Alabama, and named the vehicle "Ingenuity." Known during planning stages as the Mars Helicopter Scout or simply the Mars Helicopter, the nickname "Ginny" later entered use in parallel to the parent rover Perseverance being affectionately referred to as "Percy."

6.1.4 Technical Specifications

Ingenuity's dimensions are as follows:

- Fuselage (body): 13.6 × 19.5 × 16.3 cm (5.4 × 7.7 × 6.4 in).
- Landing legs: 0.384 m (1 ft 3.1 in).
- Rotor diameter: 1.2 m (4 ft).
- Height: 0.49 m (1 ft 7 in).
- Initial mass:
    - Batteries: 274 gm (9.6 oz).
    - Total 1.8 kg (4.0 lb).
- Batteries: 273 gm (9.6 oz).
- Power: 350 watts.

Its flight characteristics are as follows:

- Rotor speed: 2,400 rpm.
- Blade tip speed: <0.7 Mach.
- Originally planned operational time: 1–5 flights within 30 sols.
- Endurance: up to 167 seconds (per flight).
- Maximum range, flight: 625 m (2,050 ft).
- Maximum range, radio: 1,000 m (3,300 ft).
- Maximum planned altitude: 12 m (39 ft).
- Maximum possible speed:
    - Horizontal: 10 m/s (33 ft/s).
    - Vertical: 3 m/s (9.8 ft/s).
- Battery capacity: 35–40 Wh (130–140 kJ).

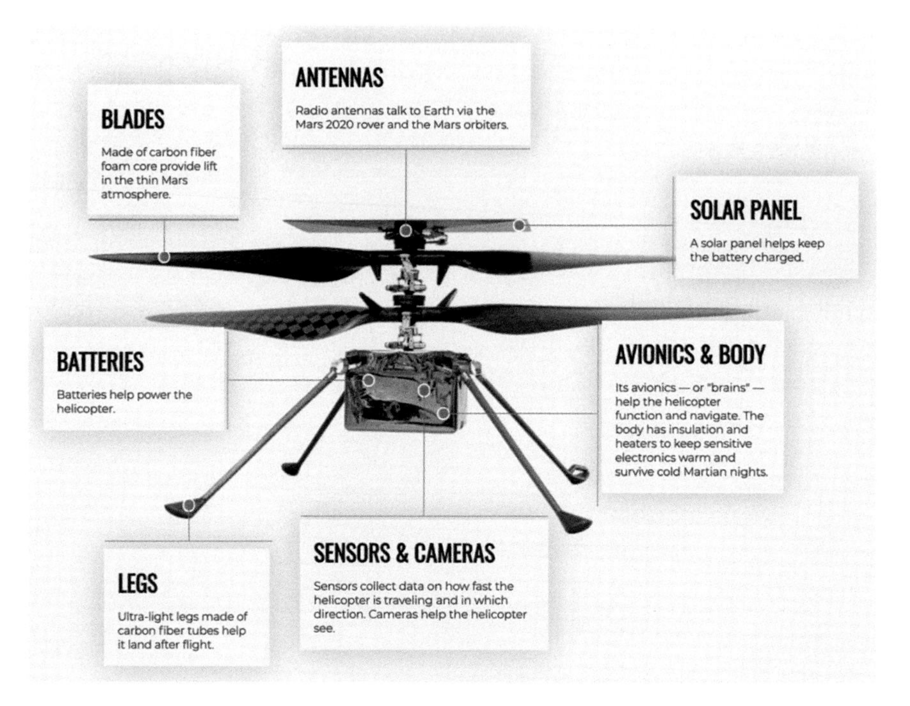

Fig. 6.1 Anatomy of Ingenuity. Photo courtesy of NASA/JPL-CalTech

6.1.5 Systems

Ingenuity carries two cameras, a downward-looking black-and-white navigation camera (NAV) and a color camera to capture terrain images for return to Earth. Although it is an aircraft, to endure the acceleration and vibrations during launch it was constructed to spacecraft specifications. It also includes radiation-resistant systems capable of operating in the environment of Mars. The inconsistent Mars magnetic field precludes the use of a compass for navigation, so Ingenuity relies on various sensors grouped in two assemblies. All of the sensors are commercial off-the-shelf units.

The Upper Sensor Assembly, with an associated vibration isolation elements is mounted on the mast, close to the center of mass of the vehicle to minimize the effects of angular rates and accelerations. It consists of a cellphone grade Bosch BMI-160 Inertial Measurement Unit (IMU) along with a Murata SCA100T-D02 inclinometer that is used only on the ground prior to flight to calibrate the IMU accelerometers biases.

The Lower Sensor Assembly consists of a Garmin LIDAR Lite v3 altimeter, the two cameras and a secondary IMU, all of which were mounted directly onto the Electronics Core Module rather than on the mast. The down-facing Omnivision OV7251 camera supports "visual odometry" in which images are processed for navigation solutions that enable the position, velocity and attitude of the vehicle to be calculated, plus other variables.

The helicopter uses solar panels to recharge its six Sony lithium-ion cells with a capacity of 35–40 Wh. Flight duration is not constrained by the available power, but by the motors heating up one degree centigrade every second.

It uses a Qualcomm Snapdragon 801 processor with the Linux operating system. Among other tasks, this processor controls the visual navigation algorithm via a velocity estimate derived from terrain features tracked by the navigation camera. The Qualcomm processor is connected to two flight-control microcontroller units to perform the necessary flight-control functions.

Communications between Ingenuity and Perseverance is via a system with two identical radios using monopole antennas. The radio link makes use of the low-power Zigbee communication protocols which are implemented using 914 MHz SiFlex 02 chipsets mounted in both the rover and helicopter. The communication system will communicate equally in all directions and is designed to relay data at 250 Kbps over distances of up to 1 km (3,280 ft).

Fig. 6.2 An Ingenuity systems test in the space simulator. Photo courtesy of Photo courtesy of NASA/JPL-CalTech

6.2 TECHNOLOGY DEMONSTRATION FLIGHTS

The original plan called for a 30-day technology demonstration period in which Ingenuity would fly up to five times at heights of 3–5 m (10–16 ft) for up to 90 seconds each. However, the expected lateral range was exceeded while making the third flight, and the duration was exceeded by the fourth flight. With these technical successes, Ingenuity had achieved its original objectives. The flights proved the helicopter's ability to fly in the extremely thin air of another planet over a hundred million miles from Earth without direct human control, running semi-autonomously and performing maneuvers that were planned, scripted and transmitted to it by JPL.

6.2.1 Deploying the Helicopter

Before Ingenuity could take flight, it was positioned squarely in the middle of "Wright Brothers Field," a 10 x 10 m (33 x 33 ft) patch of Martian real estate selected for its flatness and lack of obstructions. Once the helicopter and rover teams verified that Perseverance was where they wanted it to be, the elaborate process of deploying the helicopter began.

"As with everything involving the helicopter, this type of deployment has never been done before," said Farah Alibay, Mars Helicopter Integration Lead for the Perseverance rover. "Once we start the deployment there is no turning back. All activities are closely coordinated, irreversible, and dependent on each other. If there is even a hint that something is not going as expected, we might decide to hold off for a sol or more until we have a better idea what is going on."

The helicopter deployment began on April 3, 2021 and was spread over six sols. On the first sol the team activated a bolt-breaking device that released a locking mechanism which had held the helicopter tight against Perseverance's belly for launch and Mars landing. The next sol, they fired a cable-cutting pyrotechnic device to permit the mechanized arm that held Ingenuity to start to rotate the helicopter out of its horizontal position. This, in turn, enabled the rotorcraft to extend two of its four landing legs.

During the third sol of the deployment sequence, a small electric motor finished rotating Ingenuity until it latched, completely vertical. On the fourth sol the two remaining legs moved into position.

On each of those four sols, the WATSON imager took confirmation imagery of Ingenuity as it incrementally unfolded until it was "hanging" 13 cm (5 in) above the ground. At that point, the helicopter was connected to the rover by a single bolt and a couple of dozen tiny electrical contacts. On the fifth sol the team used the final opportunity for Perseverance to ensure that Ingenuity's six battery cells were fully charged.

Fig. 6.3 Ingenuity with four legs. Photo courtesy of NASA/JPL-CalTech

On the sixth (and final) scheduled sol of the deployment phase, the team needed to confirm three things: (1) that Ingenuity's four legs were firmly on the ground, (2) that the rover drove around 5 m (16 ft) away, and (3) that both vehicles were communicating via their onboard radios. This milestone also initiated the 30-sol clock during which all preflight checks and flight tests were scheduled to occur. As it turned out, Ingenuity performed so well that the 30-sol limit was lifted and the testing continued.

As with deployment, the helicopter and rover teams approached the upcoming flight test program methodically. Ingenuity survived the first cold night of the solo period on the surface. On April 9, the team began further tests, including wiggling the rotor blades and verifying the performance of the IMU, as well as testing the entire rotor system during a spin-up to 2,537 rpm while Ingenuity's landing gear remained firmly on the ground.

6.2.2 The First Flight Test on Mars

With the deployment and preflight tests completed, the engineers began getting ready to attempt the first flight. Perseverance received and relayed the final data from JPL mission controllers to Ingenuity. Several factors determined the precise time for the flight, including modeling of local wind patterns plus measurements provided by the Mars Environmental Dynamics Analyzer (MEDA) aboard the rover.

The flight was initially scheduled for April 11, but was postponed after a final preflight test when the helicopter's rotors were spun up to full speed and then aborted by the expiry of a "watchdog" timer that was to oversee the command sequence and halt activities in the event of a problem. After some analysis, the team simply modified the timing of commands slightly. Ingenuity successfully performed the high-speed rotor test on April 16th. On April 19th it spun up to 2,537 rpm, lifted off, ascended at ~1 m/sec (3 ft/sec) and hovered in place 3 m (9.8 ft) above the surface for 40 seconds. While hovering it turned through 96 degrees in a planned maneuver. The landing drew to a close the first powered controlled flight by any aircraft on another planet.

Fig. 6.4 While Ingenuity was hovering about 3 m above the Martian surface on April 19, 2021 during the historic first powered controlled flight by any aircraft on another planet, its navigation camera, which was autonomously tracking the ground, took this image. Photo courtesy of NASA/JPL-CalTech

For a video of this historic flight, go to: https://youtu.be/wMnOo2zcjXA

Several hours after the flight, Perseverance downlinked Ingenuity's first set of engineering data, images and video from the rover's NavCams and Mastcam-Z. From this data, the Mars Helicopter team determined their first attempt to fly at Mars was a success.

On the next sol all the remaining engineering data collected during the flight, as well as some low-resolution black-and-white imagery from the helicopter's own NavCam was transmitted. On the third sol of this phase the two images taken by the helicopter's high-resolution color camera arrived. The team used all available data to determine when and how to move forward with their next test.

6.2.3 The Transition Phase

Having proven that powered controlled flight is feasible on Mars, the Ingenuity experiment embarked on a new operations demonstration phase, exploring how aerial scouting and other functions could benefit future exploration of Mars and other worlds. This phase began after Ingenuity had flown twice more on April 22nd and 25th. The decision to add an operations demonstration was a result of Perseverance being ahead of schedule with checking out all of its systems since its landing and its science team choosing a nearby patch of Jezero Crater for its first phase of exploration. With Ingenuity's energy, telecommunications and in-flight navigation systems performing above expectation, an opportunity arose to allow it to continue exploring its capabilities without significantly impacting the rover's scheduling.

Ingenuity was now in a transitional phase that included its next two flights. The fourth on April 30th sent the rotorcraft about 133 m (436 ft) south to take aerial imagery of a potential new landing zone, prior to returning to Wright Brothers Field. This 266 m (873 ft) roundtrip effort would surpass the range, speed, and duration marks achieved on the third flight. The fifth flight on May 7th sent the helicopter on a one-way mission and a landing at a new site, named Airfield B.

6.3 THE OPERATIONAL PHASE

6.3.1 Flights from May to September 2021

The transition from a technology demonstration to an operations demonstration brought with it an expanded flight envelope. Along with one-way flights, there was more precision maneuvering, more use of its aerial-observation capabilities and more risk overall.

This phase began on May 22 with Ingenuity's sixth flight, and was followed by the seventh flight on June 8th. Meanwhile, Perseverance kicked off the science phase of its mission by leaving the Octavia E. Butler landing site.

During the first few weeks of Perseverance's first science campaign it made its way to a low-lying scenic overlook and surveyed some of the oldest geological features in Jezero Crater. While Perseverance was traveling to its first sampling area, called Crater Floor Fractured Rough, Ingenuity flew its 7th and 8th flights. On the 9th flight on July 5 it flew above the sandy ripples of the Séítah area, an area too risky for the rover to drive across, thereby demonstrating the capability for a helicopter to scout for a rover.

While Perseverance was at its first sampling site in late July and early August, Ingenuity made its 10th and 11th flight. After the rover had completed the first drilling, the helicopter made its 12th flight in order to scout out "South Séítah."

During its 13th flight on September 4th it provided a 3-D view of a mound that was strewn with rocks. The plan for this reconnaissance mission into the South Séítah area was to observe a geological target named "Faillefeu" (in honor of a medieval abbey in the French Alps) with color pictures from a height of 8 m (26 ft). By September 5, 2021 Ingenuity had completed the demonstration phase and its mission was extended indefinitely.

6.3.2 Atmospheric Pressure and Rotor Problem

In mid-September 2021, Ingenuity faced a new challenge. Atmospheric density can fluctuate over time and with seasonal changes. The fact that the density was dropping significantly affected Ingenuity's ability to get off the ground and climb. The solution was to spin the rotors faster, but there was a problem.

Here is an explanation (slightly edited) by Jaakko Karras, the Ingenuity Deputy Operations Lead:

> … we were getting ready to begin flying with a higher rotor speed to compensate for decreasing atmospheric density caused by seasonal changes on Mars. Increasing the rotor speed is a significant change to how we've been flying thus far, so we wanted to proceed forward carefully. Step one was to perform a high-speed spin test at 2,800 rpm on the ground and, if everything went well, step two was to perform a short-duration flight, briefly hovering over our current location, with a 2,700 rpm rotor speed.
>
> The high-speed spin test was completed successfully on September 15, 2021. Ingenuity's motors spun the rotors up to 2,800 rpm, briefly held that speed, and then spun the rotors back down to a stop, all exactly as sequenced for the test. All other subsystems performed flawlessly. Of particular interest was determining whether the higher rotor speeds cause resonances (vibrations) in Ingenuity's structure. Resonances are a common challenge in aerial rotorcraft and can cause problems with sensing and control, and they can also lead to mechanical damage. Fortunately, the data from this latest high-speed spin showed no resonances at the higher rotor rpm's. The successful high-speed spin was an exciting achievement for Ingenuity and gave us the 'green light' to proceed to a test flight with a 2,700 rpm rotor speed.
>
> The test flight was scheduled for September 18 (sol 206) and was to be a brief hover flight at 5 m (16 ft) altitude with a 2,700 rpm rotor speed. It turned out to be an uneventful flight – because Ingenuity decided to not take off!
>
> Ingenuity detected an anomaly in two of the small flight-control servo motors (or 'servos') during its automatic preflight checkout and did exactly what it was supposed to do, it canceled the flight.

Ingenuity controls its position and orientation during flight by adjusting the pitch of each of the four rotor blades as they spin around the mast. Blade pitch is adjusted through a swashplate mechanism, which is actuated by servos. Each rotor has its own independently controlled swashplate, and each swashplate is actuated by three servos, so Ingenuity has six servos in total. The servo motors are much smaller than the motors that spin the rotors, but they do a tremendous amount of work and are critical to stable, controlled flight. Because of their criticality, Ingenuity performs an automated check on the servos before every flight. This self-test drives the six servos through a sequence of steps over their range of motion and verifies that they reach their commanded positions after each step. We affectionately refer to the Ingenuity servo self-test as the 'servo wiggle'. The data from the anomalous servo wiggle shows that two of the upper rotor swashplate servos, servos 1 and 2, began to oscillate with an amplitude of approximately 1 degree about their commanded positions just after the second step of the sequence. The software detected this oscillation and promptly canceled the self-test and flight.

Our team is still looking into the anomaly. To gather more data, we had Ingenuity execute additional servo wiggle tests during the past week, with one wiggle test on September 21 and one on September 23. Both tests ran successfully, so the issue isn't entirely repeatable.

One theory for what is happening is that moving parts in the servo gearboxes and swashplate linkages are starting to show some wear now that Ingenuity has made well over twice as many flights as we originally planned (thirteen completed versus five planned). Wear in these moving parts would cause increased clearances and increased looseness, and could explain servo oscillation. Another theory is that the high-speed spin test left the upper rotor at a position that loads servos 1 and 2 in a unique, oscillation-inducing way that we haven't encountered before. We've a number of tools available for working through the anomaly and we're optimistic that we'll get past it and back to flying again soon.

Our team will have a few weeks of time to complete our analysis because Mars will be in solar conjunction until mid-October and we won't be uplinking any command sequences to Ingenuity during that time. Ingenuity will not be completely idle during this time, though. Ingenuity and Perseverance will be configured to keep each other company by communicating roughly once a week, with Ingenuity sending basic system health information to Perseverance. We will receive this data on Earth once we come out of conjunction. It will tell us how Ingenuity performs over an extended period of relative inactivity on Mars.

For a 3:43 minute video of the test, go to: https://www.youtube.com/watch?v=BBvhMqleHYk

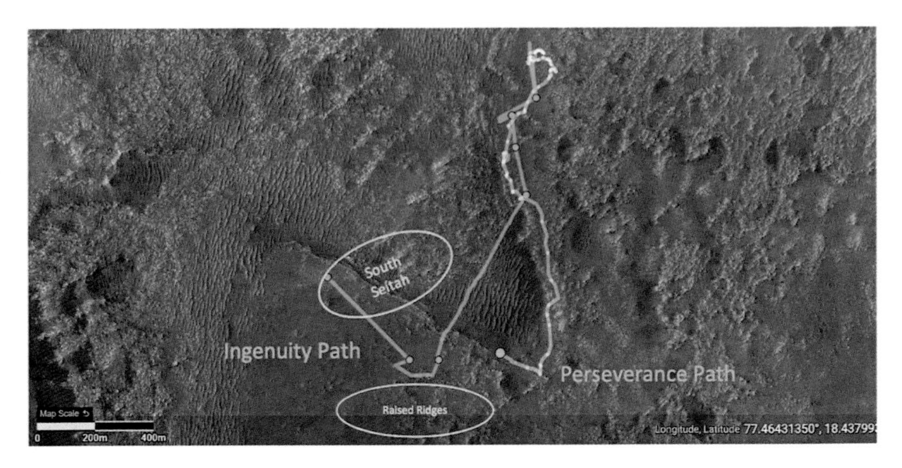

Fig. 6.5 Ingenuity's flight path. Photo courtesy of NASA/JPL-CalTech

6.4 FLIGHT LOG

Flight	Sol	Date 2021	Horizontal Distance		Maximum Altitude		Maximum Groundspeed		Duration
#			m	ft	m	ft	m/s	mph	sec
1	58	4/19	0	0	3	~10	0	0	39.1
2	61	4/22	4	~13	5	~16	0.5	~1	51.9
3	64	4/25	100	~328	5	~16	2	~4.5	80.3
4	69	4/30	266	~873	5	~16	3.5	~8	116.9
5	76	5/7	129	~423	10	~33	2	~4.5	108.2
6	91	5/22	215	~705	10	~33	4	~9	139.9
7	107	6/8	106	~348	10	~33	4	~9	62.8
8	120	6/21	160	~525	10	~33	4	~9	77.4
9	133	7/5	625	~2,051	10	~33	5	~11	166.4
10	152	7/24	233	~764	12	~39	5	~11	165.4
11	163	8/4	383	~1,257	12	~39	5	~11	130.9
12	174	8/16	450	~1,476	10	~33	4.3	~10	169.5
13	193	9/4	210	~689	8	~26	3.3	~7	160.5
14	241	10/24	2	~6.5	5	~16	0.5	~1	23.0
15	254	11/6	407	~1,335	12	~39	5	~11	128.8
16	268	11/21	116	~381	10	~33	1.5	~3	107.9
17	282	12/5	187	~614	10	~33	n/a	n/a	177
18	Last flight for 2021 to the northern edge of South Séítah is scheduled for late December.								

Through flight 17, Ingenuity had flown for a total duration of 31 min and 45 sec covering ~3.6 km (~2.25 mi).

6.5 FUTURE HELICOPTERS

Ingenuity has proven that flight on Mars is practical, and can contribute tangible scientific value even for a vehicle that was not designed to do much in the way of science at all. The question now is how its spectacular performance will influence NASA's future Mars exploration strategy.

As it happens NASA was thinking about this since long before Ingenuity reached Mars. About 3 years ago, roboticists at JPL and AeroVironment got together and started to sketch out what a next generation Mars helicopter might look like. How would a helicopter effectively scale up? What kinds of science instruments could it carry? And which missions could only be attempted using such a helicopter?

The result was the 30 kg (66 lb) Mars Science Helicopter (MSH) or "hexacopter" capable of undertaking unique science without requiring the support of a rover.

As Bob Balaram, Ingenuity Chief Engineer at JPL and one of the authors of a white paper on the Mars Science Helicopter explains, the first element in JPL's design approach was to provide as many options as practicable to the scientific community. That meant thinking about all kinds of different vehicle sizes and mission architectures. "Ingenuity could be scaled both up and down. We could make it even smaller to serve as a scout. Or we could scale it up into a full-size standalone helicopter. And of course there are things in between in the 5 kg (11 lb) class, involving taking samples from distant sites and flying them back to a lander for analysis."

JPL presented this menu of Mars helicopter options to planetary scientists and asked them to imagine what kinds of new research might be possible with each platform. And there's a lot to imagine.

"This is a brand new way of looking at Mars," Balaram says. "Aerial mobility gives you reach, range, and resolution. You can reach places that no wheeled vehicle can travel to. You can travel kilometers every day. And depending on what altitude you fly at, you can get whatever resolution you want using your instruments. We were just telling the scientists, think big!"

Currently, NASA's Mars Exploration Program is focused on sample return, and while a helicopter could play a unique and compelling role in such a mission it's by no means the obvious choice. Unless NASA decides that helicopters on Mars are absolutely the way to go, and funds MSH directly, the next Mars helicopter will have to survive a competitive proposal process that weighs potential science against cost, complexity, and risk.

As of right now (Balaram says) the MSH concept is mature enough for a broad range of potential science missions, therefore the next step is to optimize it for a specific mission scenario that takes into account a landing location, time of year, and overall goals and constraints.

Fig. 6.6 "Hexacopter". Photo courtesy of NASA/JPL-CalTech

Given that Ingenuity has been "very effective, I'm hopeful that NASA will give us the chance to engage with Mars in a completely new way with MSH. We've opened up aerial mobility on Mars. We have landed in a few places and flown a little bit here and there. So let's put our imaginations to the test and see what we could do if we had access to the whole planet. What could we achieve? That's a challenge is for all of us. To imagine that, and then make it happen."

6.6 THE INGENUITY TEAM

As Project Manager, MiMi Aung assembled a multidisciplinary team of scientists, engineers, and technicians leveraging all of NASA's expertise. The JPL team was never larger than 65 full-time-equivalent employees, but those at AeroVironment and NASA's Ames and Langley Research Centers brought the total to 150. They included:

MiMi Aung	Ingenuity Mars Helicopter Project Manager
Bob Balaram	Chief Engineer
Teddy Tzanetos	Operations Lead

Håvard Fjær Grip	Chief Pilot
Josh Ravich	Mechanical Engineering Lead
Nacer Chahat	Senior antenna/microwave engineer

On June 15, 2021 the Space Foundation named the team behind Ingenuity the 2021 winner of the John L. "Jack" Swigert, Jr. Award for Space Exploration.

IMAGE LINKS

Fig. 6.1 https://upload.wikimedia.org/wikipedia/commons/e/e8/Anatomy_of_the_Mars_Helicopter.png

Fig. 6.2 https://caltech-prod.s3.amazonaws.com/main/images/PIA23153-mars-helicopter.original.jpg

Fig. 6.3 https://upload.wikimedia.org/wikipedia/commons/d/de/Ingenuity_Helicopter_with_fully_deployed_legs_%28cropped%29.png

Fig. 6.4 https://www.nasa.gov/sites/default/files/styles/full_width/public/thumbnails/image/ncam_flight10000000.pbin_.jpeg?itok=Wd_LuAQe

Fig. 6.5 https://mars.nasa.gov/system/resources/detail_files/26134_PIA24797-Flight12.jpg

Fig. 6.6 https://spectrum.ieee.org/media-library/eyJhbGciOiJIUzI1NiIsInR5cCI6IkpXVCJ9.eyJpbWFnZSI6Imh0dHBzOi8vYXNzZXRzLnJibC5tcy8yNzE0NzI2My9vcmlnaW4uanBnIiwiZXhwaXJlc19hdCI6MTY2NzQ5OTE2NX0.o8X5ZC6alAWSS5iZRkJ7cY7bXhBajxBJZxZcwMZ_AWA/image.jpg?width=980

7

Mars 2020 Science and Engineering Teams

The Mars 2020 Perseverance team is made up of scientists and engineers from a variety of disciplines, with international participation from organizations around the world and principal investigators from the US, Italy, France, Spain, Norway and Canada.

7.1 NASA HEADQUARTERS

Within the NASA Headquarters organization, many directorates and divisions support the programs, projects and fundamental research conducted by its field centers nationwide. The Mars 2020 Program was primarily run by JPL, which is administered for NASA by the California Institute of Technology (CalTech). To understand the Mars Exploration Program, the reader requires to understand the Headquarters organization first, and then those of the supporting organizations. Basically, Headquarters funds the various programs and projects and provides management to the other organizations. Essentially, once a program like Mars 2020 is defined, key management and scientific support personnel are brought into a Program Office from the mainline organizations described below.

There are five Directorates within NASA Headquarters:

- Science Mission Directorate (SMD).
- Space Technology Mission Directorate (STMD).
- Exploration Systems Development Mission Directorate (ESDMD).
- Space Operations Mission Directorate (SOMD).
- Aeronautics Research Mission Directorate (ARMD).

M. von Ehrenfried, *Perseverance and the Mars 2020 Mission*,
Springer Praxis Books, https://doi.org/10.1007/978-3-030-92118-7_7

On September 21, 2021, NASA Administrator Bill Nelson announced that the Human Exploration and Operations Mission Directorate (HEOMD) would be divided into the Exploration Systems Development Mission Directorate that would be responsible for developing programs for the agency's Artemis lunar exploration initiative and future Mars exploration, and the Space Operations Mission Directorate to run the International Space Station and low Earth orbit commercialization efforts. This change was prompted by the increasing space operations in low orbit and development programs that were well underway to undertake deep space exploration. It would ensure that these critical areas had focused oversight. With two areas focused on human spaceflight, one mission directorate could operate in space while the other was developing future space systems. This reorganization did not change the roles and missions of NASA field centers.

Associate Administrator Kathy Lueders, who was put in charge of HEOMD in June 2020, would run the Space Operations Directorate and NASA named Jim Free to lead Exploration Systems Development. Free is a former Director of the Glenn Research Center who also served as a Deputy Associate Administrator in HEOMD.

While there are aspects of each that relate in some way to the Mars 2020 mission, it is the Science Mission Directorate that most directly supports the Mars mission. Typically, a program organization taps into the mainline organization for people and resources.

The Headquarters people most prominent on television and blogs concerning the on-going Perseverance rover and Ingenuity helicopter activities are:

- Bill Nelson, NASA Administrator.
- Dr. Thomas Zurbuchen, Associate Administrator for SMD.
- Dr. Lori Glaze, Director of SMD's Planetary Science Division.
- Eric Ianson, Director of the Mars Exploration Program in addition to his existing role as Deputy Director of Planetary Sciences Division.
- Dr. Michael Meyer, Lead Scientist for MEP and the Mars Sample Return.
- George Tahu, Acting Deputy Director and Lead Program Executive for the Mars Exploration Program.
- Dr. Mitch Schulte, Program Scientist for the Mars Exploration Program.
- Dave Lavery, Program Executive for the Ingenuity Mars Helicopter.
- Jeffery Gramling, Director of the Mars Sample Return Program.

They works directly with the various JPL organizations and the Mars 2020 team described below.

7.1.1 Science Mission Directorate

Dr. Thomas Zurbuchen is the NASA Associate Administrator for the Science Mission Directorate (SMD). He is responsible for directing and overseeing the nation's space research programs in Earth and space sciences. The Directorate engages the external and internal science community to identify and prioritize science questions in an effort to expand the frontiers of five broad pursuits:

- Earth Science.
- Planetary Science.
- Biological and Physical Sciences.
- Heliophysics.
- Astrophysics.

By a variety of robotic observatory and exploration vehicles, and by sponsored research the Directorate provides virtual human access to the farthest reaches of space and time, as well as practical information about changes on our planet. It engages the nation's scientists, sponsors scientific research, and develops and deploys satellites and probes in collaboration with NASA's partners around the world to answer fundamental questions requiring the view from and into space. SMD endeavors to understand the origin, evolution and destiny of the universe and the nature of the strange phenomena that shape it.

7.1.2 Space Technology Missions Directorate

The Space Technology Missions Directorate (STMD) develops transformative space technologies to enable future missions. Technology drives exploration to the Moon, Mars and beyond. STMD is focused on advancing technologies and testing new capabilities at the Moon that will be critical for human missions to Mars. In many ways, the Moon will serve as a technology testbed and proving ground for Mars. Technology has advanced since the Curiosity rover landed on Mars in 2012. These advances were infused into the design of the Perseverance rover for Mars 2020, as well as for the helicopter technology demonstration by Ingenuity.

By engaging and inspiring entrepreneurs, researchers and innovators STMD is creating a community of America's best and brightest working on the nation's toughest challenges. Space technology research and development take place at NASA field centers, universities and national laboratories. STMD can leverage partnerships with other government agencies, as well as with commercial and international partners.

Investments in space technologies provide solutions both on Earth and in space. NASA technology turns up in nearly every corner of modern life. NASA makes space technologies available to commercial companies so that they can provide real world benefits.

7.1.3 Explorations Systems Development Mission Directorate

The Explorations Systems Development Mission Directorate (ESDMD) is to define and manage systems development for programs critical to Artemis and develop an integrated plan for exploration of the Moon and Mars.

7.1.4 Aeronautics Research Missions Directorate

Research by the Aeronautics Research Missions Directorate (ARMD) directly benefits today's air transportation system, the aviation industry, and the many passengers and businesses that rely upon aviation every day. ARMD scientists, engineers, programmers, test pilots, facilities managers and strategic planners focus on aviation's future. They design, develop and test technologies that will make aviation much more environmentally friendly, uphold safety in ever more crowded skies, and ultimately transform the way we fly. Its long range plan for aeronautical research is aimed at the next 25 years and beyond.

NASA's aeronautics research is primarily done at Ames Research Center in California, Armstrong Flight Research Center in California, Glenn Research Center in Ohio, and Langley Research Center in Virginia.

7.1.5 Mars Exploration Program

In 1994 NASA announced the Mars Exploration Program (MEP), which was initially known as the Mars Surveyor Program, and assigned the lead role for implementation to JPL. MEP explores Mars to support NASA Headquarters' Science Mission Directorate (SMD). It currently has a number of rovers and orbiters, contributes to Mars missions conducted by national and international partners, and is formulating and developing various future orbiter and surface missions. Organized programmatically, MEP missions mutually support each other in three key aspects:

- Scientific discoveries by one mission can drive the formulation of the scientific focus of future missions. Acquired data provides essential information for the execution of future missions. For example, orbiter missions provide data for characterizing and certifying the landing site candidates for future missions, while assets on the surface give ground truth and detailed context for orbital remote-sensing results.
- Observations by assets on the surface are interpreted and placed into a broader context using results from other areas of the planet, helping to inform our understanding of Mars as a whole.
- A mission might develop and demonstrate engineering capabilities that make feasible future missions (e.g. entry-descent-landing capabilities). MEP orbiters enhance the scientific return of landed missions by serving as communications relays.

7.1.6 Space Operations Mission Directorate

The Space Operations Mission Directorate (SOMD) that was carved out of the previous Human Exploration and Operations Directorate is in charge of human spaceflight operations on the International Space Station and commercialization of low orbit.

7.2 JET PROPULSION LABORATORY

JPL is a Federally Funded Research and Development Center (FFRDC) which is managed and operated by CalTech under a NASA contract. Its primary role is the construction and operation of planetary robotic spacecraft, though it also conducts Earth-orbit and astronomy missions. It is also responsible for operating the NASA Deep Space Network with facilities around the world. It has a staff of over 6,000, plus another thousand contractors.

For more than half a century NASA's fleet of increasingly sophisticated orbiters, landers and rovers has helped transform science fiction into reality, in particular enabling them to piece together detailed information about Mars. These robotic explorers have revealed details of the Red Planet's diverse surface, atmosphere, and weather, and about the presence of water in its past and its potential for life through the ages and for astronauts in the future.

JPL has played a pivotal role in deep space exploration, starting when the first successful mission, Mariner 4, conducted the first Mars flyby in 1965, and the landings on the planet by the Viking missions in 1976. The era of Mars rovers opened when Mars Pathfinder landed in 1997 and deployed its Sojourner rover. JPL has designed, built and operated all five of the successful Mars rovers. Like NASA Headquarters, it is organized into directorates and into specific programs such as the Mars 2020 mission. The directorates supply programs with specific expertise and people, in some cases more directly than others.

7.2.1 Engineering and Science Directorate

Charles Whetsel, Director of the Engineering and Science Directorate, says the way to understand JPL is to understand that it is a "matrix organization" where the senior management and new mission definition staff represent a fairly thin "project/product" function, and the vast majority of the talent for each project, such as for the Mars 2020 mission, is staffed from the Engineering and Science Directorate. The Project Manager, Flight System Manager and Payload Manager roles are staffed from the Mars Directorate, but the chief engineer, as well as the mechanical, telecom, thermal, etc. subsystem delivery teams are all staffed from the "discipline-oriented" divisions of the Engineering and Science Directorate. The technical divisions – Mechanical Systems Divisions, Instrument Systems Division, and so on – also provide the talent and facilities to deliver products to JPL's other planetary and Earth science missions and instruments, as well as for astrophysics and fundamental physics and for Deep Space Network operations.

7.2.2 Planetary Science Directorate

According to Dr. Robert Braun, the Planetary Science Directorate (PSD) was created in March 2020 by a merger of the former directorates for Solar System Exploration and Mars Exploration. Around half of JPL's budget and workforce are associated with the PSD's projects, which include Perseverance, Ingenuity, Europa Clipper, Psyche, VERITAS, Mars Sample Return, Lunar Trailblazer and numerous missions in operation across the solar system. In a memorandum to its staff Braun stated:

> This organizational change is designed to improve alignment and communications with NASA HQ and our partner organizations, improve the integration and communication of our priorities and challenges across the Lab and strengthen our collaboration and interactions with the planetary science community. We will team with other organizations across the Lab in the development of a strong, diverse and inclusive personnel pipeline for the future roles in the Directorate and across the Lab. We will learn from each other and be an even greater force for discovery going forward.
>
> Planetary science has evolved into a broad and growing enterprise with the addition of new scientific disciplines and questions, an increasing breadth of organizations and nations engaged, and the expanding engineering complexity of missions undertaken. The newly-formed PSD will play to our strengths by formulating, developing and operating planetary science missions and instruments that only JPL can. With a science-first mentality, we will work to deliver and operate each planetary science project within the required technical, cost and schedule constraints.

7.2.3 Astronomy and Physics Directorate

Mr. Keyur Patel is the JPL Director of the Astronomy and Physics Directorate, which studies the universe and our place in it. In addition to the formation and evolution of stars and galaxies, it studies the diversity of planets orbiting other stars. Like the Mars 2020 program this organization is also looking for signs of life but much farther away, in other star systems.

It defines, acquires, formulates, implements and operates activities associated with astronomy and fundamental physics by:

- Providing support to the science community, both within JPL and outside, in areas aligned with the Directorate's program and projects.
- Advocating and advancing associated technologies both from NASA and other sponsors.
- In addition to being responsible for the implementation of applicable portions of the JPL Strategic Plan, it supports the development and execution of strategic plans by other agencies.

7.2.4 Earth Science and Technology Directorate

Dr. James Graf is the Director of the Earth Science and Technology Directorate, which focuses on the interconnected components of the Earth – the atmosphere, hydrosphere, lithosphere, biosphere, and cryosphere – and how these complex, dynamic systems work together to produce an integrated whole. Studying how Earth is changing enables scientists to better understand, predict and respond to natural hazards, weather, climate, and freshwater requirements. To address the challenges, JPL pioneers space-based observations made practicable by unique, innovative technologies. Specific areas of study are:

- Understanding our water cycle and monitoring freshwater availability.
- Measuring sea level rise and quantifying contributing ocean and other Earth system processes.
- Monitoring the solid Earth to understand and respond to natural hazards.
- Understanding our carbon cycle and changing ecosystems.
- Enabling improvements in weather and air quality forecasts.
- Aligning research and missions with the 2017–2027 Decadal Survey for Earth Science and Applications from Space, which recommended a rich and complex set of Earth science and application measurements which would address the challenges and opportunities associated with the five strategic themes listed above.

7.2.5 Interplanetary Network Directorate

The Interplanetary Network Directorate is JPL's programmatic focal point for deep space communications, navigation, and mission operations, and performs premier solar system science and astrophysics. It is responsible for the design, development, operation and services for three of NASA's "mission enabling" systems: the Deep Space Network (DSN), Advanced Multi-Mission Operations System (AMMOS), and Planetary Data System (PDS) support nodes.

Five orbiters currently make up the Mars Relay Network to transmit commands from Earth to surface missions and receive science data back from them. While some commands and telemetry can be sent directly to/from Earth, for the most part the huge quantities of science data collected by rovers and landers cannot, because it would take too long. Most data traveling back to Earth must first be sent to the orbiters overhead which then use their more powerful transmitters to send it to antennas on Earth, including those of the Deep Space Network.

Although the Mars Relay Network has expanded to include more spacecraft and more international partners, and currently comprises NASA's MRO, MAVEN, Odyssey and ESA's TGO, with ESA's Mars Express orbiter being available for emergencies, every new surface mission adds complexity to the relay sessions.

Curiosity and InSight are near enough to each other on Mars that they are almost always visible by the orbiters at the same time when they fly over. Perseverance landed far enough away that it can't simultaneously be seen by MRO, TGO, and Odyssey but sometimes MAVEN, which has a larger orbit, will be able to see all three vehicles at the same time. Since JPL uses the same set of frequencies when communicating with all three orbiters, they have to carefully schedule when each orbiter talks to each lander. The Mars Relay Network achieves new throughput records when it returns Perseverance's huge datasets.

7.3 INTERNATIONAL SUPPORT

7.3.1 ESA Support to Mars 2020

To communicate with Earth from its landing site in Jezero Crater, Perseverance relies on spacecraft orbiting Mars to relay the imagery and other data it collects back to Earth and pass on the commands from engineers beamed across space in the other direction via the Deep Space Network operated by JPL and the Estrack network operated by ESA.

International support for the landing of Perseverance on Mars was provided by Mars Express, present since 2003, and the ESA/Roscosmos ExoMars Trace Gas Orbiter (TGO), which took up orbit in 2016. TGO relayed important data from Perseverance to Earth a mere 4 hours after landing. Mars Express is monitoring the local conditions at Jezero Crater.

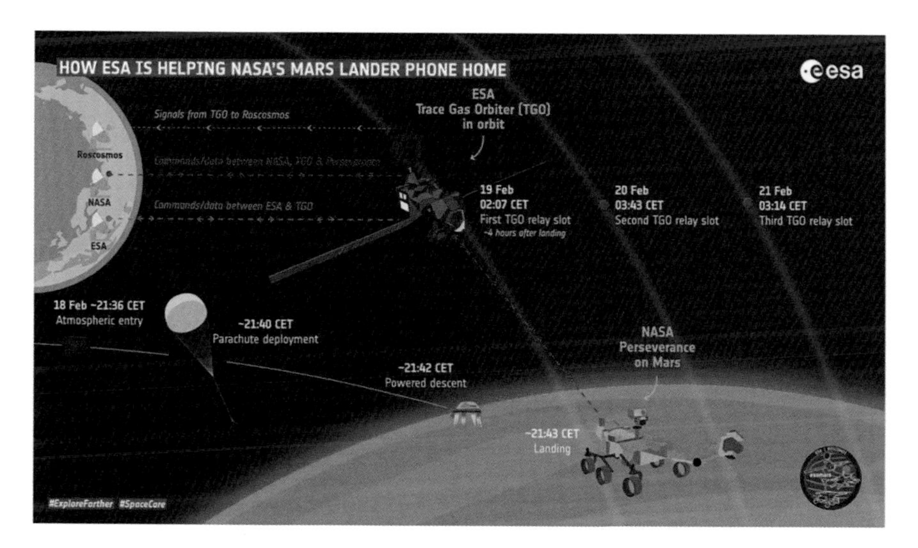

Fig. 7.1 How ESA supports Perseverance. Photo courtesy of ESA

7.3.2 ESA Support to the Mars Sample Return Mission

NASA and ESA are investigating mission concepts for an international Mars Sample Return campaign between 2020 and 2030. Three launches would be necessary to accomplish landing, collecting, storing and finding samples and delivering them to Earth.

NASA's Mars 2020 mission is to store a set of rigorously documented samples in canisters in strategic areas to be retrieved by a later mission for delivery to Earth. Two missions are envisioned for this pioneering next step:

- A NASA launch will send the Sample Retrieval Lander mission to place a platform near the Mars 2020 site that is to deploy a small ESA rover, the Sample Fetch Rover, to drive to retrieve the cached samples. Once it has collected them in what can be likened to an interplanetary treasure hunt it will return to the lander platform and load the samples into a single large canister on the Mars Ascent Vehicle (MAV). It will make the first ascent from Mars and then release the container in orbit.
- ESA's Earth Return Orbiter will be timed to capture the basketball-size sample container orbiting Mars. The samples will then be sealed in the bio-containment system to prevent contaminating Earth with unsterilized material. The spacecraft will then return to the vicinity of Earth, where it releases an entry capsule carrying the sample container. As in the case of the Apollo lunar samples, the Martian samples will be safely transported to a specialized handling facility.

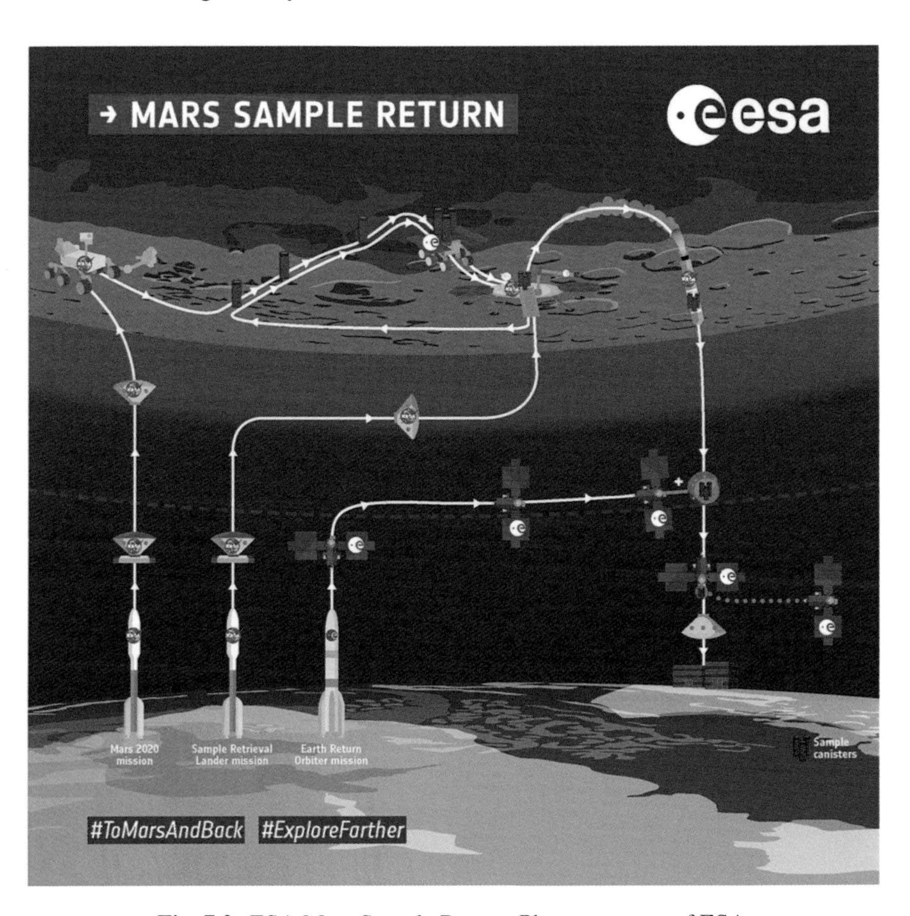

Fig. 7.2 ESA Mars Sample Return. Photo courtesy of ESA

This proposal is more thoroughly discussed in Chapter 8.

7.3.3 Support to Other Mars Exploration Missions

NASA contributed a pair of Electra radio communications systems to the ESA ExoMars Trace Gas Orbiter and assisted with the navigation, tracking and data return of that spacecraft. In addition, NASA contributed the mass spectrometer and the primary electronics for the Mars Organic Molecule Analyzer (MOMA) instrument of the ExoMars rover mission to be launched in 2022 that will allow detection and characterization of organic molecules on Mars.

NASA also assisted with navigation and tracking of the Indian Space Research Organization's Mars Orbiter Mission (MOM), now safely in orbit around Mars. The two agencies have an agreement to work on future Mars missions.

NASA also has an agreement to assist the United Arab Emirates with the Hope mission that was launched in 2020 and went into orbit around Mars in 2021. The Mohammed Bin Rashid Space Centre led the design, development and operation of the mission.

NASA is planning to help develop one of the remote sensing instruments for the Martian Moon Exploration (MMX) mission that is currently being developed by the Japan Aerospace Exploration Agency (JAXA) for launch in 2024.

IMAGE LINKS

Fig. 7.1 https://www.esa.int/var/esa/storage/images/esa_multimedia/images/2021/02/how_esa_is_helping_nasa_s_mars_lander_phone_home/23146587-1-eng-GB/How_ESA_is_Helping_NASA_s_Mars_lander_phone_home_article.png

Fig. 7.2 https://www.esa.int/var/esa/storage/images/esa_multimedia/images/2019/05/mars_sample_return_overview_infographic/19412751-6-eng-GB/Mars_Sample_Return_overview_infographic_article.jpg

8

Mars Sample Return

8.1 THE OVERALL PLAN

The Mars Sample Return (MSR) campaign seeks to bring samples of Martian rocks and soil back to Earth, where they can be investigated in unprecedented detail using all the capabilities of terrestrial laboratories. It is part of NASA's Mars Exploration Program, a long-term effort of robotic exploration of the Red Planet. NASA is collaborating with ESA to develop the advanced technologies and hardware required for the campaign.

With the Perseverance rover already collecting samples, the MSR management team is structured to support the program. While the funding and programmatic details are not all in place, the management is as follows:

- Jeff Gramling from NASA Headquarters is the Program Director.
- Dr. Michael Meyer, also from NASA Headquarters is the Mars Lead Scientist.
- Dr. Richard Cook with JPL is the MSR Project Manager as of March, 2021 (a position previously held by Dr. Robert D. Braun).

The NASA/ESA Mars Sample Return Science Planning Group (MSPG) feeds input to the MSR managers and has developed a Science Management Plan that addresses the scientific and technical issues and proposes a working list of high-level requirements for the Sample Return Facility (see Section 8.8), providing a timeline of key decision points.

A JPL & JSC team visited eighteen BioSafety Level-4 (the highest level) and contamination-controlled facilities to appreciate the challenges of conducting science with returned Martian samples. Fifteen NASA/ESA Returned Sample

M. von Ehrenfried, *Perseverance and the Mars 2020 Mission*,
Springer Praxis Books, https://doi.org/10.1007/978-3-030-92118-7_8

Scientists have been competitively chosen for the Perseverance Science Team. These teams participate in Sample Caching Strategy Workshops that provide input to the management and science teams.

Overview

The Mars Sample Return mission architecture is designed to continue the work begun by NASA's Perseverance rover, which is collecting samples and sealing them in tubes to be cached at a yet-to-be-selected location in Jezero Crater for a future mission to retrieve and transport to Earth.

The current schedule is for NASA to launch the Mars Sample Return Lander in 2026 for arrival at Mars in 2028, touching down close to the where Perseverance cached the samples. In fact, if Perseverance is still operational at that time then it could directly deliver additional samples to the Mars Ascent Vehicle.

8.2 SCIENCE OBJECTIVES

Mars Sample Return (MSR) has long been designated as a high priority by the international planetary exploration community. Analysis of returned samples is expected to enormously advance our understanding of the planet's history and evolution, in particular concerning its past and present habitability.

NASA and ESA signed a joint Statement of Intent in 2018 to further define the roles and responsibilities for leading respective elements of the potential MSR flight campaign. Contributions from other international agencies are also being considered. One key challenge in developing a science program is to give every partner fair opportunity to participate in the scientific discovery process. To that end, NASA/ESA jointly chartered the MSR Science Planning Group (MSPG) to help to develop an overarching science management strategy that will guarantee fair and competitive balance for all international stakeholders.

8.2.1 Science Management Framework

In developing the Framework, the MSPG relied upon five cornerstone tenets to guide its deliberations:

- Transparency: Access to samples must be fair and the processes which define sample access must be as transparent as possible.
- Science maximization: Management of scientific and sample-related processes must optimize the scientific productivity of the samples via careful selection of science investigations and consideration of sample preservation.

- Accessibility: Throughout the MSR process international scientists must have multiple opportunities to participate in a variety of capacities.
- Return on investment: Agencies providing the investments to undertake the MSR campaign should receive demonstrable benefits.
- One Return Canister/One Collection: The returned samples ought to be managed as a single collection even if the samples are physically housed in different facilities and the ownership of sample shouldn't be pro-rated according to investment.

This was a key first step towards developing the Science Management Plan that will guide the overall MSR science program.

The MSPG developed an overarching strategy detailing the high level structures, decision making bodies, and processes required to implement the program, and highlighted near-term actions which can be taken following official approval of the MSR campaign.

The Framework is organized into three categories of activities through which the science community can participate in the MSR process:

1 Management and management planning: Beginning in 2020, these are the entities and processes that are involved in the oversight of returned sample science, and will offer guidance for operational functions such as curation and planetary protection.
2 Planning for facilities: Also starting in 2020, groups will assess scientific considerations pertinent to the design and functional requirements of the various facilities necessary to implement MSR.
3 Returned sample science processes: Beginning in 2026, these bodies will define, provide sample access for and execute the essential investigations that will drive the objectives of the MSR campaign.

8.2.2 Specific Objectives

Several groups including the MEPAG and MSPG have refined the MSR science objectives during the past decade. These are continually being examined by the scientific community, and include:

1 Determine the chemical, mineralogical, and isotopic composition of the crustal reservoirs of carbon, nitrogen, sulfur, and other elements with which they have interacted, and characterize the C-, N-, and S-bearing phases down to submicron spatial scales in order to document processes that could sustain habitable environments on the planet today and in the past.
2 Evaluate the evidence for prebiotic processes, past life, and/or extant life on Mars by characterizing the signatures of these phenomena in terms of their structure/morphology, biominerals, organic molecular and isotopic compositions, and other evidence within their geological contexts.

3 Interpret the conditions of water-rock interactions by their mineralogical products.
4 Constrain absolute ages of major crustal geological processes, including sedimentation, alteration and diagenesis, volcanism/plutonism, regolith formation, weathering, and cratering.
5 Understand paleo-environments and the history of near-surface water by characterizing the sedimentary sequences and their clastic and chemical components, depositional processes and post-depositional events.
6 Constrain the mechanisms and timing of the process of planetary accretion and differentiation, and the subsequent evolution of the crust, mantle, and core.
7 Determine how regolith is produced and modified, with its temporal and spatial variations.
8 Characterize risks to human explorers from biohazards, material toxicity, and dust/granular materials, and also contribute to assessments of in-situ resources in support of human presence.
9 Determine the preservation potential for the chemical signatures of extant life and prebiotic chemistry in the present surface and shallow subsurface by evaluating the oxidation state as a function of depth, permeability, and other factors.
10 Constrain the early composition of the Martian atmosphere, the rates and processes of atmospheric loss/gain over geological time, and the rates and processes of atmospheric exchange with surface condensed species.
11 Determine the age, geochemistry and conditions of formation and ensuing evolution of climate-modulated polar deposits by detailed examination of the compositions of H_2O, CO_2 and dust constituents, and the fine structure of the stratigraphy.

8.3 SAMPLE RETRIEVAL LANDER

The first element of the MSR campaign is a NASA-led Sample Retrieval Lander (SRL) that will be responsible for retrieving the Perseverance-collected samples, loading them into an Orbiting Sample (OS) container and then launching the OS into a stable Mars orbit by a Mars Ascent Vehicle (MAV). Next, the ESA Earth Return Orbiter (ERO) will rendezvous with the OS, capture it, and then return it to Earth.

The SRL payloads would include the Sample Fetch Rover (SFR) with its Sample Transfer Arm (STA) and the NASA-provided OS and MAV. It would notionally launch on a Type III/IV trajectory timed to arrive near the beginning of the Mars spring in the northern hemisphere. The longer Earth-Mars transfer time for SRL allows the Earth Return Orbiter mission, launched in the same window, to arrive in time to relay communications for SRL. The solar-powered SRL would carry out its surface activities during local spring and summer, maximizing the power available, and then before the start of autumn in the northern hemisphere launch the retrieved samples into orbit. This timing was to ensure the surface activities would be finished prior to a significant decrease in available solar power and in advance of the potential for global dust storms.

The Entry, Descent and Landing (EDL) procedures for SRL will exploit heritage from the Perseverance mission, in particular hypersonic guided entry, supersonic parachute deployment and the use of terrain-relative navigation improve landing safety and reduce landing dispersion. A sky crane will lower the legged lander to the surface.

Fig. 8.1 Lowering the Sample Retrieval Lander onto Mars. Photo courtesy of NASA/JPL-CalTech

Throughout the surface mission of SRL, ERO will provide communication relay services to SRL, SFR and Perseverance. It will be supported by any relay assets from the current Mars Relay Network that are still operational at that time.

Fig. 8.2 The SRL with its SFR and the MAV. Photo courtesy of NASA/JPL-CalTech

8.4 ESA'S SAMPLE FETCH ROVER

Given that Perseverance will be collecting samples in several areas, the cache(s) must be left at location(s) where another vehicle can find them, collect them and transfer them back to the Sample Retrieval Lander for loading into the container aboard the Mars Ascent Vehicle.

Once on Mars, the ~120 kg (265 lb) Sample Fetch Rover will egress from SRL and set off to retrieve samples that were cached by Perseverance at one or more depots. The currently envisioned surface mission timeline allocates 150 sols for SFR to complete this retrieval and return to SRL. With a nominal 4 km (2.5 mi) landing dispersion for SRL, this mandates a requirement for high levels of SFR autonomy and mobility, with support for drive ranges of up to 500 m (1,640 ft) per sol and also efficient operations during tube retrieval at the depot site(s) and subsequent tube transfer to SRL.

For a short video on how the Sample Fetch Rover will pick up samples, go to: https://dlmultimedia.esa.int/download/public/videos/2020/02/031/2002_031_BR_001.mp4

On returning to SRL, the Sample Transfer Arm of the SFR will load the sample tubes into the container atop the MAV. Once a full load has been transferred, the MAV will launch and deploy into Mars orbit a basketball-size container holding the samples.

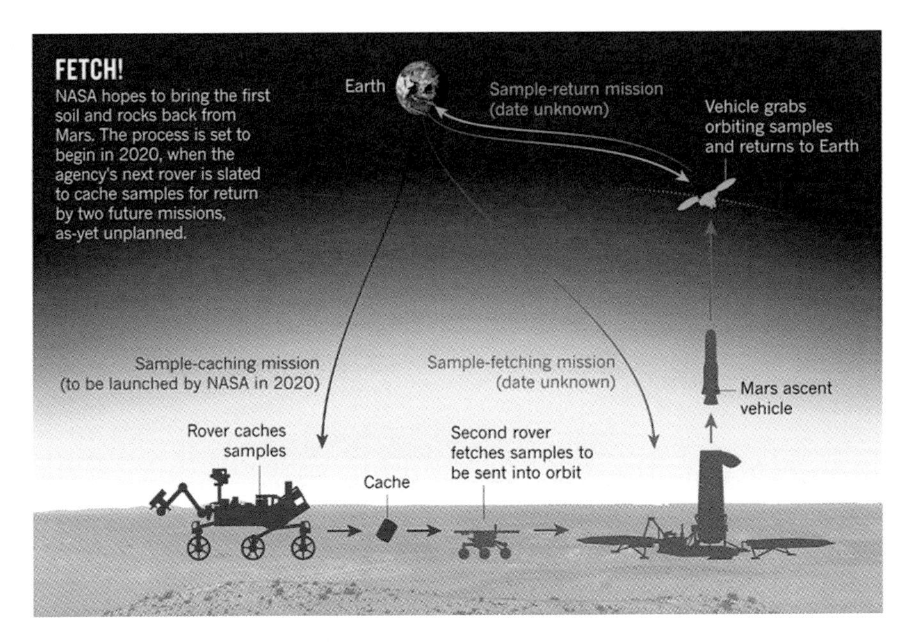

Fig. 8.3 The Fetch concept. Photo courtesy of NASA/JPL-CalTech

Fig. 8.4 The Sample Robotic Arm transfers the sample tubes from SFR to SRL. Photo courtesy of NASA/JPL-CalTech

Sophisticated algorithms to spot the sample tubes cached on the Martian surface have already been developed by an industry effort led by Airbus. The dedicated robotic arm with a grasping unit to pick up the tubes is being designed by a pool of European industries.

SFR will be roughly the size of NASA's Opportunity rover. ESA's contractor is Airbus Defence and Space. It will travel typically 200 m (650 ft) per sol over a period of 6 months to find and retrieve up to 36 sample tubes and return them to SRL. The retrieval rover will have only four wheels in order to save on mass and complexity. The type, size and number of wheels has been chosen to better cope with the known terrain and with the speed and performance needed to reach the depot(s) and return to SRL in a timely manner.

8.5 MARS ASCENT VEHICLE

Once the Orbiting Sample (OS) container is aboard the MAV, that will launch into a stable Mars orbit; current plans target a circular orbit at an altitude of 350 km (217 mi). SRL/MAV launch operations must be carefully coordinated with ERO to ensure that the orbiter has a line-of-sight to the lander for critical event communication and tracking during launch. The accuracy of the orbit achieved determines the complexity of ERO operations to locate and rendezvous with the passive OS. The desired accuracies correspond to better than 10 km (6.2 mi) in semi-major axis and less than 1 degree of divergence in inclination.

Fig. 8.5 The Orbiting Sample container. Photo courtesy of NASA/JPL-CalTech

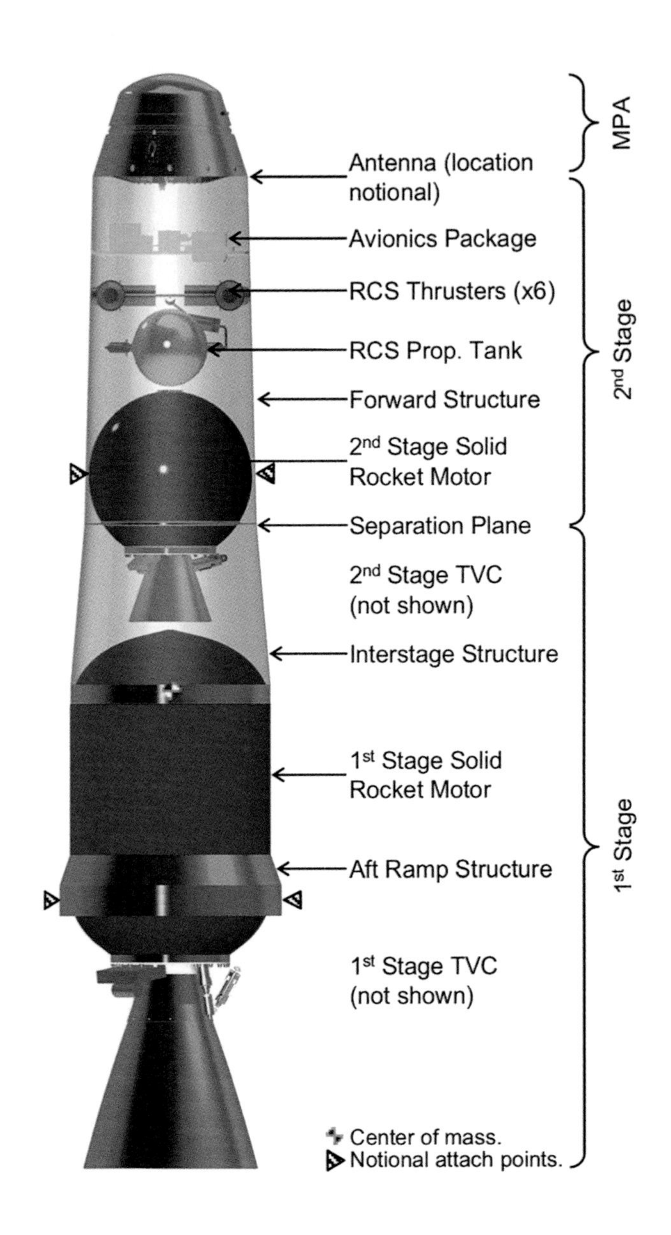

Fig. 8.6 The MAV design concept. Photo courtesy of NASA/MSFC

Because Martian gravity is just 38% that of Earth, a rocket designed to launch a payload into orbit can be much smaller. Furthermore, the MAV requires only to carry a payload of 14–16 kg (30–35 lb). Based on preliminary design constraints, the MAV would be no taller than 2.8 m (9.2 ft) and no wider than 57 cm (1.9 ft) with a total liftoff mass not exceeding 400 kg (881 lb).

Engineers at the NASA Marshall Space Flight Center (MSFC) in Alabama first considered a single-stage hybrid propulsion system which would have burned a solid wax-based fuel with a liquid oxidizer and they teamed up with two hybrid propulsion providers to perform test firings, but in 2020 NASA decided to use a two-stage MAV powered by solid rocket motors. Northrop Grumman Systems will provide propulsion support and products for the MAV development, under contract to MSFC.

Fig. 8.7 The MAV lifts off from Mars carrying the OS container. Photo courtesy of NASA/MSFC

8.6 EARTH RETURN ORBITER

The current schedule calls for the launch of the ESA Earth Return Orbiter (ERO) along with the Earth Entry Vehicle (EEV) in 2026 for arrival in 2028 prior to the launch of the Mars Ascent Vehicle. It will use technological heritage from ESA's

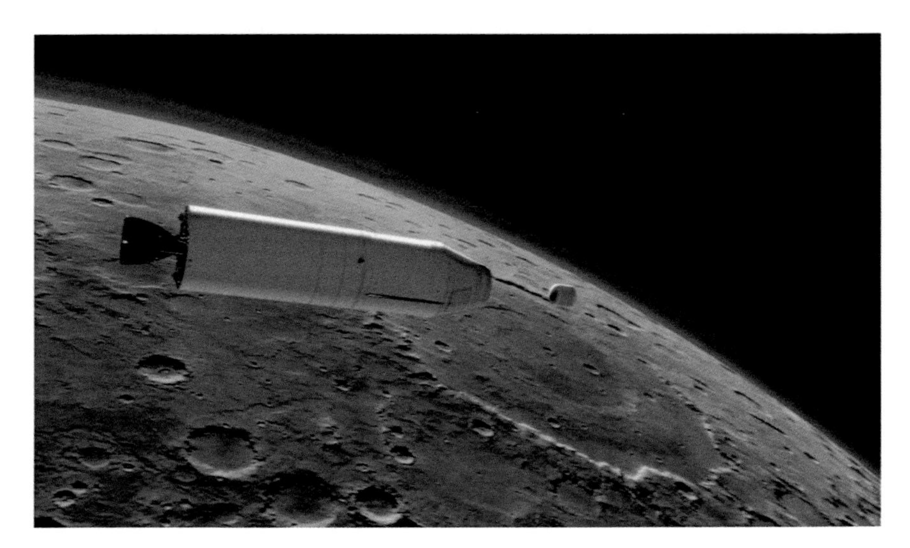

Fig. 8.8 The MAV deploys the OS container in Mars orbit. Photo courtesy of NASA/MSFC

most recently launched science mission, BepiColombo; a joint venture between ESA and JAXA conducted under ESA's leadership. Both use electric propulsion and multi-stage detachable modules. To catch the Orbiting Sample it will exploit heritage in autonomous rendezvous from the Automated Transfer Vehicles with which ESA delivered cargo, fuel and oxygen to the International Space Station.

The five year mission of ERO will begin with an Ariane rocket launching from French Guiana. Once in orbit around Mars it will act as a communications relay for the surface missions of the Mars Sample Return campaign.

In 2020 ESA selected Airbus as prime contractor for the ERO spacecraft, taking advantage of the company's experience in autonomous rendezvous and docking, flight-proven technologies and decades of interplanetary space optical navigation expertise, including developing the JUICE spacecraft as Europe's first mission to Jupiter.

With an initial mass of 6,000 kg (13,277 lb), the 6 m (20 ft) tall ERO spacecraft will have solar arrays spanning more than 40 m (131 ft) with an area of 144 m^2 (1,550 ft^2); among the largest ever created. Solar electric propulsion will be used for the year-long voyage to Mars, followed by chemical propulsion for insertion into Mars orbit. The electric propulsion will be used to spiral down to a targeted circular rendezvous orbit at an altitude of approximately 400 km (250 mi).

In the most critical phase of its mission, ERO will detect, rendezvous with and capture the basketball-size Orbiting Sample (OS) released in orbit by the MAV. An onboard Capture, Containment, and Return System will isolate the Orbiting Sample, then transfer it within the ERO to an Earth Entry Vehicle. Following a year-long return trip the Earth Entry Vehicle will be deployed on a trajectory to enter Earth's atmosphere and touch down at a pre-defined landing site, possibly the Utah Test and Training Range.

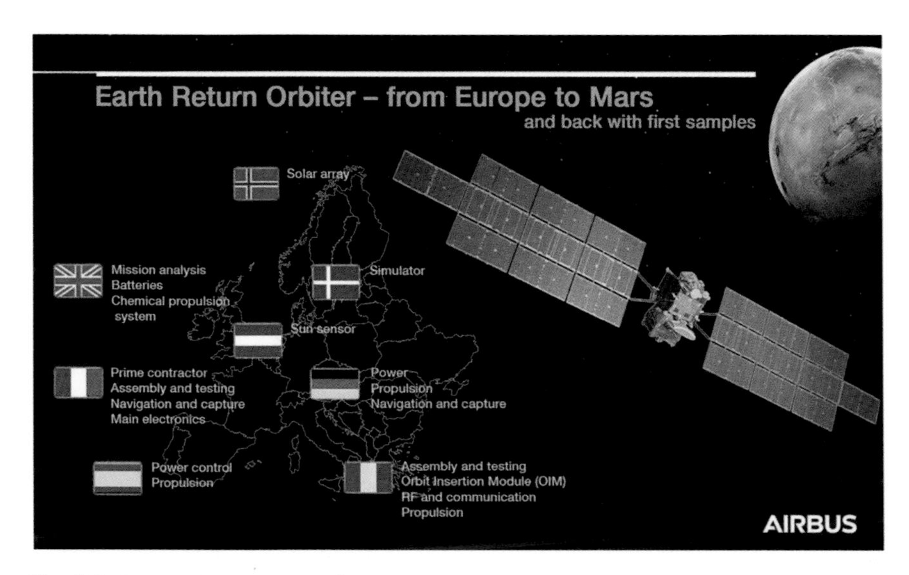

Fig. 8.9 International support for ERO. Photo courtesy of Airbus

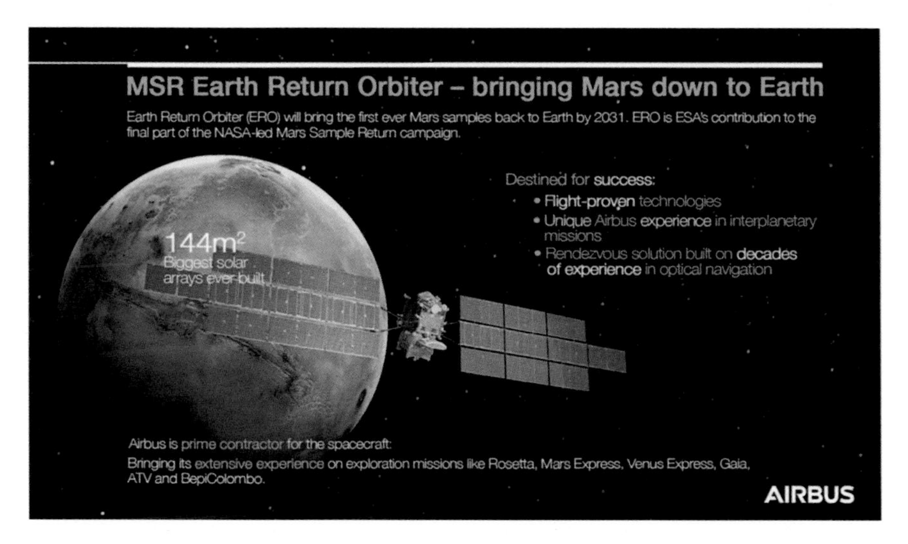

Fig. 8.10 Bringing samples of Mars down to Earth. Photo courtesy of Airbus

8.7 EARTH ENTRY VEHICLE

8.7.1 Organization

In 2018, NASA created a new multi-center Earth Entry Vehicle (EEV) team to mature concepts and prepare for a potential 2026 Mars Sample Return mission. The new team is run out of the Mars Exploration Directorate at JPL with strong support from the NASA Langley and Ames Research Centers. It works together in many areas:

- Concept development and maturation.
- Mission architecture and systems engineering.
- Engineering analysis and testing.
- Risk mitigation and hardware certification planning.
- NASA/ESA joint MSR studies.
- Identification and closure of critical technology gaps.
- Planning and budgeting for potential flight implementation.

The 2018 study objectives were to:

1 Demonstrate that feasible EEV design solutions can be developed which meet key vehicle requirements, fit within the MSR architecture, and are programmatically realistic.
2 Down-select to two concepts that would carry forward into the FY 2019 design maturation and testing.
3 Formulate FY 2019–2020 risk reduction, technology development, and conceptual design work needed to achieve a successful Orbiter Mission Concept Review (MCR) in early 2020.

8.7.2 Concept

The overall goal of NASA's Mars Sample Return project is to bring surface and atmosphere samples to Earth for detailed analysis. The NASA Langley Research Center's Earth Entry Vehicle performs a critical role in the mission, protecting the sample container from reentry heating and deceleration loads during atmospheric entry, descent, and landing.

The basic design of the EEV was developed in the 1998–2001 timeframe for the 2003–2005 MSR project which was canceled in 2001 but this work was followed by technology development efforts through 2004. Mission studies in 2004 called for an EEV system validation flight test in 2010 and the launch of MSR in 2013. New development studies were underway at JPL and Langley at the time.

Since mission plans do not call for sterilizing the samples before returning them to Earth, sample containment must be maintained for protection of the terrestrial environment. In a draft requirement for the 2003–2005 MSR project, the NASA Planetary Protection Officer required the probability of a particle larger than 0.2 microns being released into Earth's biosphere must be less than one in a million. This constraint requires the MSR EEV to operate with a reliability never before imposed upon a planetary entry vehicle. Probabilistic risk assessment techniques were used to quantify the reliability of the design options, and these drove many aspects of the mission and vehicle design in order to achieve that high reliability requirement.

8.7.3 Flight Plan

The Earth Return Orbiter will carry the EEV back toward Earth on a near-miss trajectory. A deflection maneuver to change from a biased trajectory to nominal Earth-entry trajectory will be performed several days prior to releasing the EEV. The spacecraft will then use a second deflection maneuver to fly past Earth. The EEV will be spin-stabilized in an orientation for a zero degree angle of attack on penetrating the atmosphere.

The EEV will make a ballistic entry which imposes a high heat and deceleration, a method used by the entry vehicles of the Genesis and Stardust missions. After slowing to a terminal velocity of 41–45 m/sec (92–101 mi/hr) it will land in soft soil. The yet-to-be-selected landing site should have controlled ground and air-space across a wide area of soft terrain, with the Utah Test and Training Range being the obvious candidate. After recovery, the EEV and the sample container within it will be transported to a dedicated handling facility.

The entry speed intended for the EEV of the 2003–2005 MSR project was 11.56 km/s (25,858 mi/hr). An entry flight path angle of 25 degrees was chosen in order to limit the stagnation point reentry heat flux to the level achievable in ground test facilities, so that the thermal protection materials could see adequate qualification testing prior to flight. These trajectory values will need to be updated for a future mission.

8.7.4 The EEV Design

The baseline design of the MSR EEV is a blunt body 0.9 m (3 ft) in diameter with an entry mass of 44 kg (97 lb). Inside this will be the 16 cm (6.3 in) spherical OS container that will accommodate 0.5 kg (1.1 lb) of Martian material. Its forebody is a 60 degree (half-angle) cone with a spherical nose; the aft side is concave with a central hemispherical lid that latches shut after the OS container is inserted. The JPL-designed sample container fits in the center of the EEV, inside a 5 mm (0.2 in) thick flexible containment vessel that is sealed in Mars orbit prior to heading home. The multiple, dissimilar layers of containment are to protect Earth against common-cause failure modes.

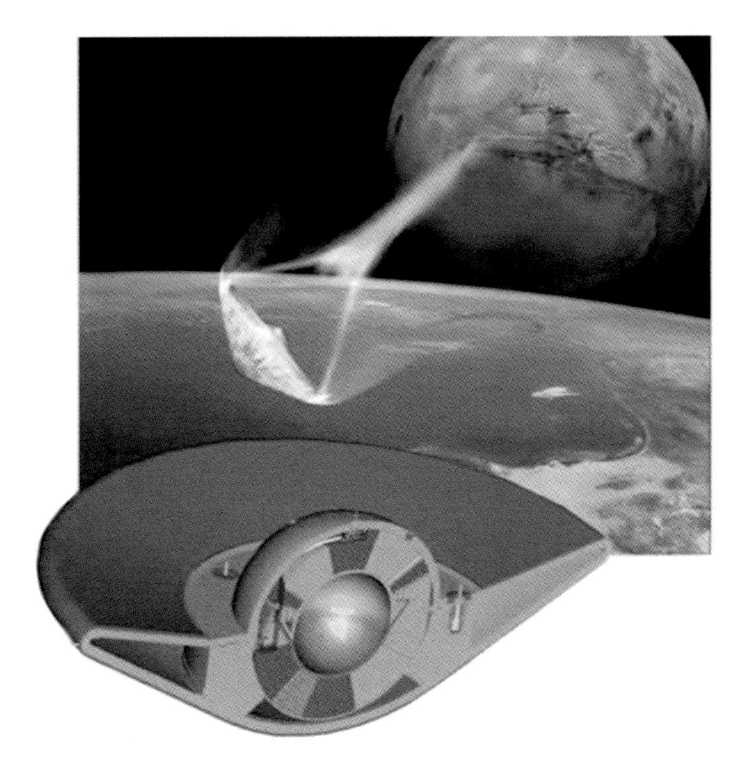

Fig. 8.11 The EEV penetrating the atmosphere. Photo courtesy of NASA Langley Research Center

The extremely high reliability requirement, combined with data on the low-but-finite failure rates of spaceflight hardware, led to the elimination of active flight systems from the EEV. It has no attitude control system; it is simply released by the ERO on a ballistic, free-fall trajectory, and is spin-stabilized to maintain the proper orientation. It was specifically designed such that its aerodynamics would act as a passive backup to ensure desired orientation during reentry. Simulations show that even if a failure of the spacecraft's release system spin-stabilizes the EEV at an angle of attack of 180 degrees (i.e. fully backwards) the vehicle will flip itself over to nose-forward in the hypersonic flow regime prior to the entry heat pulse. There is likewise no parachute. Instead, the vehicle is designed for a terminal velocity landing onto soft terrain. Tests at potential landing sites have shown that the nominal soft terrain at landing sites similar to the Utah Test and Training Range will cushion landing loads below the 2,500G level necessary to preserve the scientific value of the samples.

In the event of a hard surface, the deceleration load is limited by the spherical impact energy absorber that surrounds the sample container and containment vessel. The absorber is a cellular structure, with resin-impregnated Kevlar and carbon walls braced by carbon foam in order to prevent buckling. Under high deceleration, the walls deform and tear to absorb energy as the sphere crushes. Full-velocity impacts on concrete at Langley Research Center indicate that the energy-absorbing sphere will keep the loads below the 3,500G design value for containment of the sample materials. Non-linear finite element impact analysis models and simulations of the energy absorber have been developed, validated against drop tests, and used to evaluate landing behavior for a range of ground impact conditions. By this analysis it has been demonstrated that the spherical energy absorber accommodates angled impacts arising from vehicle dynamics during terminal descent, ground slope irregularities and lateral winds.

The 2003–2005 design for the EEV envisioned a 2-D carbon-carbon structure to maintain the shape of the vehicle during the 130G entry deceleration load. This gave a high-temperature capability that reduced the risk of a structural failure in the event of overheating at the thermal protection system bond lines. However, difficulties in developing analysis methods for progressive crack growth using carbon-carbon proved sufficiently challenging that a titanium structure is now being considered.

Probabilistic risk assessment evaluation indicated that the titanium structure is a viable design, with the elimination of analysis uncertainties balancing the loss of the high-temperature capabilities of carbon-carbon. The EEV thermal protection system protects the vehicle and the sample container from the heat of entry. The reliability requirement drove the selection of the forward thermal system to fully dense carbon phenolic (FDCP) based on its extensive flight heritage. FDCP has seen thousands of tests and hundreds of flights in uses ranging from missile heat shields to solid rocket nozzle throats to the primary heat shields for the Galileo and Pioneer Venus entry probes. Two distinct types of FDCP thermal protection are required: tape-wrapped FDCP for most of the EEV forebody, and chopped-molded FDCP at the forward stagnation point where the spherical nose prevents use of the tape-wrapped material. Sample coupons of both types of FDCP have passed arc-jet tests for the predicted peak heat flux and full heat load but further development is still required as there are gaps in the information on the heritage fabrication techniques for the chopped-molded FDCP.

8.7.5 Conclusion

The design of the new EEV mission is expected to be similar to the one that was baselined for the earlier planned 2003–2005 MSR mission. The longevity of this design speaks to the robustness of the concept where aerodynamic performance, heritage materials, and passive impact attenuation form the basis of meeting the stringent sample containment requirement. But to mature the proposal for flight readiness in support of a future MSR mission the technology advancements and risk reduction activities presented here must be pursued, culminating with a full-scale EEV system validation flight test.

8.8 SAMPLE RECEIVING FACILITY

The following are excerpts from NASA-sponsored groups such as the MEPAG, MSPG, IMEWG and IMOST which have been discussing the requirements for handling sample material from Mars and indeed other planetary bodies for over 30 years. In the last decade the focus has been on the Mars 2020 mission and its follow on MSR effort. The timeframe for the samples to be returned to Earth is now forecasted for 2031.

8.8.1 The Plan

It has been widely understood for many years that an essential component of an MSR mission is a Sample Receiving Facility (SRF) to take delivery of the flight hardware that is recovered, open it, extract the sample container and the samples and conduct an agreed upon test protocol, while ensuring strict containment and contamination control of the samples. Materials that prove to be non-hazardous, or are rendered non-hazardous by sterilization, may then be transferred to long-term curation.

While the general concept of an SRF is relatively straightforward, there have been many discussions about implementation planning. The Mars Exploration Program examined the attributes of an SRF to establish its scope, including its minimum size and functionality, budgeting (capital cost, operating costs, cost profile) and development schedule. The proposed approach was to arrange for three independent design studies, each led by an architectural design firm, and then compare the results. While there were many design elements in common, there were significant differences in the way human operators were to interact with the systems. In aggregate these studies provided welcome insight into the attributes of a future SRF and the factors that would need to be considered by future programmatic planning.

8.8.2 Sample Receiving and Curation Facility

A hugely important part of an MSR campaign is the facility (or facilities) which will support the processing, basic characterization and eventual allocation of the sample material. At present, although final decisions about the return part of the mission have yet to be finalized, it is almost certain that the designated landing site of the EEV with the canister containing the samples will be in the US. This will need a sample receiving facility to be based in the landing area, as was the case for the Genesis and Stardust missions. The recovery activities, opening the canister and subsequent processing will all be governed by planetary protection requirements. The chain of what happens when and where is under current consideration.

Once we have Martian material on Earth, there will be a systematic process to evaluate the type of material available. To ensure fair access to samples by the scientific community, and also to carry out and evaluate the tests for planetary protection, it is likely that an international board will be established to oversee a preliminary examination of samples as well as their subsequent distribution. The structure and governance of the body that will oversee sample curation, analysis and distribution is under consideration.

There will undoubtedly be a number of Sample Curation Facilities (SCF), one in the US and one in Europe for sure, with the one in the US almost certainly being co-located with the SRF.

8.8.3 Tour of Existing Laboratories

During 2019 and 2020 the NASA Tiger Team RAMA (a mnemonic of authors' names) visited eighteen top-level containment biosafety laboratories and pristine space-mission facilities worldwide. This was to better understand their practices, capabilities and lessons-learned in order to aid in planning the SRF in support of MSR. They also toured a manufacturer of mobile and modular high-containment facilities, as well as manufacturers of isolators and gloveboxes. In addition, they visited ESA facilities already developing a novel double-walled isolator (DWI) and robotic handling in support of MSR. The NASA team visits covered several construction modalities, including:

- A new traditional brick-and-mortar facility.
- Use of an existing brick-and-mortar facility rated at the highest level of biological safety (Biosafety Level 4).
- A novel modular BSL-4 approach.
- A hybrid combination of brick-and mortar, modular, and existing facilities.

An MSR SRF would combine the complexities of both high-containment and pristine facilities. Although combining negative-pressure biocontainment and positive-pressure cleanroom technology would be challenging it is achievable. Furthermore, while adopting the Returned Sample Science requirements of the Mars 2020/MSR mission for contamination control (reduction of organics and bioburden) is particularly challenging for an MSR SRF, it is feasible by using novel techniques and technologies. For example, ESA has begun developing a DWI approach that may prove to be an enabling technology in providing both containment and cleanliness in conjunction with the need for a pristine facility.

Depending on the complexity, traditional brick-and-mortar BSL-4 facilities can nominally take a decade or more to design, build and commission even without unexpected delays. Because of the proposed pressure regimes for the SRF, the NASA team estimated that an MSR SRF from design to commissioning would take 8–12 years depending on construction modality. In order to allow adequate schedule reserve, they recommended starting the design definition phase for the potential MSR SRF as soon as possible.

Based on the notional MSR campaign schedule, NASA's construction options may already be time-limited, especially if the initial design phases are delayed.

Through their tours and subsequent conversations the RAMA team discovered that BSL-4 facilities can experience five or more years of delays during design, construction, and commissioning. This could represent a serious programmatic risk to MSR. Schedule delays have been caused by new requirements levied by regulatory agencies in order to minimize loss of containment risks, government funding availability/programmatics, poor design/construction practices, hiring inexperienced subcontractors, and poor community engagement. It is therefore critical that NASA begin MSR SRF community engagement as part of the site selection process and continue that effort through facility design, construction, commissioning, and ultimately the receipt of samples.

In addition, it will be important for NASA to begin engagement with regulatory agencies and science stakeholders to determine requirements before the facility design phase begins. An existing BSL-4 facility could be leveraged for at least some SRF activities; however, if anticipated contamination control and science requirements for the facility hold, many of the proposed SRF functions weren't deemed appropriate for any of the BSL-4 facilities visited by the RAMA team. Some of the challenges are providing enough laboratory space, accepting large equipment, keeping an MSR lab clean, and assuring adequate isolation to allow unsterilized samples to be safely released after biohazard assessment. To utilize any of the toured facilities, it might become necessary to de-scope MSR science goals and notional contamination control requirements. Furthermore, in view of the disparate nature of the proposed biohazard testing for MSR versus traditional terrestrial biohazard testing, the techniques and analytical equipment needed for MSR were not available in the labs visited. However, given the potential benefit of leveraging high-containment expertise/infrastructure, existing community buy-in, and possible cost and schedule savings, this could be further explored once the minimum science requirements have been better defined.

The following are the RAMA team's conclusions regarding an MSR SRF:

- A new brick-and-mortar approach used by all US BSL-4 laboratories to date. However this approach could be the most expensive modality, take the longest time to implement and pose significant programmatic risk of delay.

Given the current MSR timeline of returning a Martian sample in 2031 it is unclear whether this option is still viable.

- The use of an existing BSL-4 facility may be possible depending on the final contamination control and science requirements for the MSR SRF. Due to the internal dimensions of the labs visited and facility structural requirements, it is unlikely that any modifications could be made to the facility to meet cleanliness requirements. Furthermore, due to possible construction delays, possible capacity issues, and cross-contamination vectors, there may also be significant programmatic risks in sharing an existing facility.
- Another approach is to construct a contemporary modular facility. The modular elements would be installed in a traditional building or a shell structure. This approach has only been used for BSL-3/3Ag but seems feasible. A modular facility offers many advantages over a traditional brick-and-mortar option with lower costs, shorter design-construction-commissioning schedule, and flexibility for easier retrofits and future expansion.
- A hybrid approach of either a modular facility inside a new brick-and-mortar building or a modular and/or brick-and-mortar BSL-4 annex in conjunction with an existing BSL-4 space should be considered. This would offer the advantage of leveraging the strengths of each group's methods.

Beyond the construction of a facility, the RAMA team considered technologies for isolating and handling extraterrestrial materials. ESA has been studying and developing a range of sample-handling technologies. NASA and ESA ought to collaborate in developing such technologies in tandem with designing the SRF. The research and development investment for clean, remote manipulation and robotics at the onset of the facility design phase would be greatly beneficial to MSR.

The potential scope and challenges of the SRF are of course highly dependent upon the science, contamination control, and planetary protection requirements that are currently being defined. The RAMA team proposed regular interaction with science advisory and regulatory groups in order to gain feedback and seek answers to questions.

8.8.4 Preliminary Analyses

The SCF staff would be responsible for the curation and eventual allocation of samples to the international community. The allocation would (presumably) be decided by an international allocation committee operating under the auspices of an oversight board. There have been many studies outlining how such a facility might operate, the instruments to be installed, and the analyses to be performed. Planetary protection protocols for different areas and processes within the SCF are being discussed under the auspices of COSPAR. There are many issues that have to be addressed prior to analyzing a sample. The most recent study that has deliberated on potential operations models for an SRF is the MSPG joint ESA-NASA Working Group. The MSPG recognized two phases of examination that the samples would undergo in the first instance: basic characterization (BC) and preliminary examination (PE), with both occurring under full BSL-4 conditions.

An outline of a viable sequence of events is as follows. BC is non-invasive and non-destructive. At a minimum, this comprises photographing and weighing the sample tubes. All tubes would go through BC upon arrival at the SCF. The next phase, PE, will be more detailed, minimally invasive and non-destructive, and is likely to be carried out on a single tube at a time.

There are still huge unresolved issues of how BC and PE would take place. One concern is the dust that will cover the tubes. It is likely to permeate the canister and will have to be collected at different stages of the procedure for opening the return canister, the sample container, and the individual sample tubes.

Other issues include:

- How the headspace gases above the tubes ought to be extracted and what (if anything) should be used to repressurize the tubes.
- The desirability (or otherwise) of performing computer tomography (CT) scanning of the tubes prior to removing the sample.
- If, when and how material should be sterilized prior to distribution.
- How much sample will be assigned for PE testing and how this will it be selected, and so on.

It is apparent that whatever instruments and techniques are selected to undertake the BC and PE analyses they must generate sufficient information to characterize the samples and allow decision-making during the subsequent allocation process.

8.8.5 Instrumentation

The international community has made several attempts to determine the type of instrumentation and analysis that will be required for complete characterization of returned extraterrestrial samples. The required analyses can be divided up into the following categories:

- Morphology, structure and texture.
- Mineralogy and mineral chemistry.
- Organic components.
- Isotopic composition.

As yet undetermined is the split between which analyses will require to be done in the SCF as part of BC and PE, taking into account time-dependent considerations (e.g. disequilibrium reactions involving water activity once a tube is opened), and which analyses might be deferred until a sample has been sterilized.

No decisions have yet been reached as to which equipment or instruments would be required. It is also almost certain that samples would be processed in the SCF for allocation outside the SCF to individual investigators or groups with various specialist instruments that will acquire data beyond that acquired within the SRF. The lists of analytical instrumentation being considered is substantial.

8.8.6 Summary

Decades of observation of Mars by flyby, orbiting, landed and roving missions, complemented by data from terrestrial and space-based telescopes and Martian meteorites have created an enormous dataset of information. This has enabled a detailed picture of the planet to be assembled. Scientists have been able to infer relative chronologies for the planet's formation and its atmospheric, fluvial and volcanic histories. They have evaluated the likelihood (or not) of the evolution and survivability of life, and the suitability of the planet as a target for human exploration. Nevertheless, the science community doesn't know the age of the planet's crust or mantle, or when water flowed across the surface, or when the massive volcanic eruptions occurred. Such information can only be obtained by fetching samples to Earth. Only then will detailed studies of carefully prepared, specific components of the samples yield the absolute chronology of Mars. The return of samples from Mars is an essential next step in understanding the origin and evolution of our neighboring planet. Determining the biological capacity of Martian material will be an essential step in preparing for human exploration of the planet.

8.9 PLANETARY PROTECTION

8.9.1 International Requirements

In order to understand the Planetary Protection (PP) measures applicable to the MSR program and to the samples themselves, one must have knowledge of the definition of terms and rules used in this particular scientific discipline.

COSPAR and the Outer Space Treaty

In 1958 the International Council of Scientific Unions (ICSU) set up the ad-hoc Committee on Contamination by Extraterrestrial Exploration (CETEX) to offer advice on space exploration and related issues. In the next year this mandate was transferred to the newly established Committee on Space Research (COSPAR). Being an interdisciplinary scientific committee of the ICSU it was deemed to be the appropriate place to continue the work of CETEX. Since that time COSPAR has been an international forum to discuss such issues under the terms "planetary quarantine" and later "planetary protection," and it has formulated the COSPAR Planetary Protection Policy with associated implementation requirements as an international standard of protection against interplanetary biological and organic contamination.

The COSPAR Planetary Protection Policy, and its associated requirements, is not legally binding under international law but it is an internationally agreed standard providing implementation guidelines for compliance with Article IX of the United Nations Outer Space Treaty of 1967. State Parties to the Treaty are responsible for national space activities under Article VI of this Treaty, including the activities of governmental and non-governmental entities. It is the State that ultimately will be held responsible for wrongful acts committed by its jurisdictional subjects.

Updating the COSPAR Planetary Protection Policy, either as a response to recent discoveries or because of specific requests, is a process that involves members of the COSPAR Panel on Planetary Protection who represent (on the one hand) their national/international authority responsible for compliance with the Outer Space Treaty and (on the other hand) COSPAR Scientific Commission B (Space Studies of the Earth-Moon System, Planets and Small Bodies of the Solar System) and COSPAR Scientific Commission F (Life Sciences as Related to Space). Upon a consensus by relevant parties, the COSPAR Panel on Planetary Protection sends recommendations to the COSPAR Bureau for review and approval. The recently updated COSPAR Policy on Planetary Protection was published in August 2020.

Special Regions

COSPAR says a Special Region is "a region within which terrestrial organisms are likely to replicate" as well as "any region which is interpreted to have a high potential for the existence of extant Martian life." A Special Region is defined in terms of temperature and potential for water activity.

The Mars 2020 mission carries an RTG that naturally produces heat. This could raise temperatures, melt ice, and provide an environment in which microbial life could reproduce. The Mars 2020 mission was therefore restricted from landing at sites where scientists suspect water, brine, or water ice could be present or could be induced within 5 m (16.4 ft) of the surface, it having been reasoned that in the event of an unsuccessful landing an impacting spacecraft wouldn't go below that depth.

Water is key to life as we know it and water activity considers the availability of water and water vapor as potential support for life processes. A committee of the

Mars Exploration Program Analysis Group (MEPAG) reviewed and updated the description of Special Regions on Mars as locations where terrestrial organisms might manage to replicate. A region on Mars can be listed as Non-Special if the temperature and water availability will remain outside the threshold parameters. All other regions of the planet are designated as either Special or Uncertain. At present there are no Special Regions defined by the existence of extant Martian life.

While the Mars 2020 mission was restricted from landing in, entering or creating a Special Region on Mars, the Science Definition Team report concluded that the primary mission objective of exploring an ancient environment did not require the mission to access a Special Region. The Mars 2020 mission relies on engineering systems that are based on the Curiosity mission and those systems did not readily support the sterilization requirements for accessing a Special Region.

8.9.2 Planetary Protection Categories

There are five planetary protection categories defined by COSPAR. All Earth-return missions are Category V and a Mars Sample Return mission is Category Vb (Restricted Earth Return), the requirements for which are given as follows:

- The absolute prohibition of destructive impact upon return.
- Containment throughout the return phase of all returned hardware which directly contacted the target body or unsterilized material from the body.
- Containment of any unsterilized sample collected and returned to Earth.
- Post-mission there is a need to conduct timely studies of any unsterilized sample collected and returned to Earth under strict containment and with the most sensitive techniques. If any sign of a non-terrestrial replicating entity is found the returned sample must remain contained unless treated by an effective sterilizing procedure.

NASA is also investing in technologies to enable the safe transport of materials from Mars to Earth. These include technologies for containing samples and the sample containers for Mars materials, as well as technologies for sterilizing the exterior of the containers. Although it is not known whether life ever existed on Mars, concept studies for returning Martian material are conducted under safety standards that are as high as (or higher than) the standards used for transporting known hazardous material on Earth. Technologies currently under investigation include in-space sterilization approaches such as localized heat sterilization and plasma sterilization, as well as sealing methods like in-space metal brazing and explosive welding.

It has sometimes been the case in previous design studies that the requirements for PP have been seen to be in conflict with those of the science goals. It is now recognized that many of the analyses designed to respond to PP questions about the potentially hazardous nature of Martian materials are the same as those that would be applied to answering science questions about the presence (or absence) of biological matter and its characteristics. The benefit of the change in approach to both PP and mission science goals is that there is likely to be a more efficient and streamlined analytical sequence for the preliminary investigations, possibly with a more rapid decision about releasing samples from containment to permit scientific investigations to start.

8.10 THE NEXT TEN YEARS

Let's wrap up by looking ahead a decade or thereabouts:

- From 2020 to about 2026 the Perseverance rover caches its samples of Martian material at locations in Jezero Crater.
- Around 2026 or 2028, the Sample Return Lander and the Earth Return Orbiter are independently launched.
- A year later the Earth Return Orbiter enters orbit around Mars, ready to serve as a communications relay.
- The Sample Return Lander lands, the Sample Fetch Rover retrieves the cached samples and loads them into a special container inside the Mars Ascent Vehicle.
- The Mars Ascent Vehicle launches and successfully places the container into the proper orbit.
- The Earth Return Orbiter successfully rendezvouses with and captures the container, inserting it into the Earth Entry Vehicle.
- The Earth Return Orbiter departs Mars orbit and begins the year-long trip back to Earth.
- The Earth Entry Vehicle is spin-stabilized and released on a trajectory that will enter the atmosphere at the proper angle.
- The Earth Return Orbiter deflects its path to miss Earth.
- The Earth Entry Vehicle survives the intense reentry heating and lands at the selected site.
- It is transferred to a BSL-4 Sample Receiving Facility.
- The samples are analyzed by scientists around the world for decades (the rocks brought from the Moon a half century ago are still being analyzed). The conclusions will impact future manned and robotic missions to Mars and other planets and their moons.

If the Mars Sample Return mission is successful in bringing samples to Earth it will surely result in a major step forward for planetary science. Examination of material from the small number of samples won't tell scientists everything they want to know. Mars (like Earth) has a rich, complex history with different sites representing mere snapshots. The history of one site will be viewed in terms of the global context provided by observations and detailed local measurements at representative sites across the planet. Also, fundamental questions will remain regarding its potential for life, its geological and climate history and the forces that drove these changes.

In 2021, with Perseverance starting to collect samples, Dr. Michael Meyer, the Mars Exploration Program's Lead Scientist, said, "As the Mars Sample Return campaign moves forward and into the next decade, and as more spacefaring nations and commercial partners emerge, our understanding of the Martian environment will become broader and deeper. We are on the cusp of profound advances in deep space exploration that will initiate the shift from robotic to human exploration of Mars."

The first person to set foot on Mars has undoubtedly already been born and may well be reading this book!

IMAGE LINKS

Fig. 8.1 https://encrypted-tbn0.gstatic.com/images?q=tbn:ANd9GcRSxzvaOTYk3jzJjp9ZoOVgug5bR0NPJqkhJw&usqp=CAU

Fig. 8.2 https://d3i71xaburhd42.cloudfront.net/f852eb0ae9bf789e287a90db701f10e455bb2199/5-Figure5-1.png

Fig. 8.3 https://www.astroreality.com/wp-content/uploads/2020/06/graphic-mars2.jpg

Fig. 8.4 https://img.intelligent-aerospace.com/files/base/ebm/ias/image/2020/06/Sample_Fetch_Rover_transfer_mod_Copyright_NASA_JPL_CALTECH.5eebb60434c41.png?auto=format&fit=fill&fill=blur&w=1200&h=630

Fig. 8.5 https://mk0spaceflightnoa02a.kinstacdn.com/wp-content/uploads/2020/04/2560px-PIA23712-Mars-SampleReturn-OrbitingContainer-Concept-20200225.jpg

Fig. 8.6 https://mk0spaceflightnoa02a.kinstacdn.com/wp-content/uploads/2020/04/mav_concept1.jpg

Fig. 8.7 https://mars.nasa.gov/system/resources/detail_files/24764_PIA23496-web.jpg

Fig. 8.8 https://photojournal.jpl.nasa.gov/jpegMod/PIA23500_modest.jpg

Fig. 8.9 https://airbus-h.assetsadobe2.com/is/image/content/dam/products-and-solutions/space/space-exploration/mars/mars-sample-return/Infographic-MSR-ERO-from-Europe-to-Mars-and-back-CopyrightAirbus.png?wid=1920&fit=fit,1&qlt=85,0

Fig. 8.10 https://ttadigitalmedia.files.wordpress.com/2020/10/94988106-4c35-4cbb-b997-c4b2af7b2d35.jpg

Fig. 8.11 https://www.researchgate.net/profile/Louis-Glaab/publication/267203891/figure/fig1/AS:669431793467392@1536616385006/NASA-LaRC-MMEEV-MSR-concept_Q640.jpg

9

Conclusions

In 1975 NASA's Viking Project found a place in history when it became the first mission to land two spacecraft safely on the surface of Mars and return images of that landscape. Each lander conducted three biology experiments designed to look for possible signs of life. There was unexpected and enigmatic chemical activity in the soil, but no clear evidence of living microorganisms. It was concluded that Mars is self-sterilizing because the combination of solar ultraviolet radiation that saturates the surface, the extreme dryness of the soil, and the oxidizing nature of the soil chemistry actually prevents the formation of living organisms.

Almost a half century later, the scientific community is searching for life again, this time using far more sophisticated instruments and a roving capability. The current Mars Exploration Program started in 1994 but it took almost another 20 years to define the mission and the instruments they needed to search for life. In the meantime, NASA scrambled for the money to pay for a mission that would cost $2.9 billion with inflation, plus two years' of prime mission operations that would be a further $200 million. The spacecraft itself accounted for $2.2 billion while launch services for the Atlas V rocket were $243 million. If it lasts as long as Curiosity the operational costs will be higher but well worth it because that is when the science pays off. Despite its seemingly high cost, Perseverance is only the third most expensive mission to Mars after the Viking pair and the Curiosity rover.

Thousands of people have been working on the current Mars 2020 mission for a very long time and no matter how successful Perseverance will eventually be in collecting and caching samples, it will need the Mars Sample Return mission to bring the samples to Earth. This is a complicated, expensive, risky, challenging effort involving several interdependent elements. The current plan calls for two

M. von Ehrenfried, *Perseverance and the Mars 2020 Mission*,
Springer Praxis Books, https://doi.org/10.1007/978-3-030-92118-7_9

launches in 2026–2028, with samples being delivered to Earth in the 2031–2033 timeframe – so it will be 11–13 years after Perseverance's launch in 2020 before its samples reach Earth. If the lunar samples returned by the Apollo crews are an example, generations of scientists will be studying Mars' gift to Earth for many years to come.

What may become more obvious after reading this book is the awareness of the enormous human resource existing at the NASA Jet Propulsion Laboratory. Of course there are others from the NASA Centers, other government agencies such as the USGS and the Los Alamos National Laboratory, from our universities and our international partners. So enjoy reading Appendix 3 with brief biographies of Mars 2020 team members and their contributions to the many phases and aspects of the mission.

One can also appreciate the time it takes for a team to achieve a scientific and/or an engineering consensus regarding any chosen approach to solving a problem or answering science questions. It took many years for the scientific community to agree on the goals and objectives for the Mars 2020 mission. It took 5 years just to select Jezero Crater as the prime landing site. It took 8 years from the concept of a Mars helicopter until Ingenuity (all four amazing pounds of it) made the first flight of an aircraft on another planet on April 19, 2021.

Fortunately, the Perseverance rover had the opportunity to take advantage of the successful Curiosity design with significant technology upgrades, especially the increase in navigation accuracy during entry, descent and landing. Likewise, the improvement in the state-of-the-art for science instruments was, in some cases, orders of magnitude better than was available for Curiosity. One can extrapolate to the Mars Sample Return mission and imagine the technology we will have in that timeframe.

And there is the marvel of mechanical engineering incorporated into the Sample Caching Systems with arms drilling, sampling, sealing and stowing the Martian material and handing off one step of the process to the next. When Perseverance caches its precious samples this will make possible a new process, program, and series of flights.

There was a time, not so long ago, when many pundits thought we ought to start human missions to Mars immediately, even before going back to the Moon. But when you think how rapidly the state of the art has advanced in many scientific, engineering and robotic fields, one can reasonably conclude that it was then way too early for humans to attempt such a long journey. It is now evident that human flights to Mars are unlikely to begin until the early 2030s, if not later, allowing at least another decade of technological improvements. And the robot missions are adding to the scientific and engineering advancements that will apply to human flights. There is the MOXIE instrument that has demonstrated a way that future explorers can make oxygen from the Martian atmosphere; the MEDA weather station that will add to an understanding of the Martian atmosphere as a whole and help to study the shape and size of the dust particles in the atmosphere; and the SHERLOC instrument that is exposing small pieces of spacesuit material to see how they hold up in the harsh Martian environment. Also, now that we have seen how well Ingenuity assists Perseverance in planning ahead, we can foresee how well robots of many types will assist human explorers.

In operational terms, JPL knows how to get to Mars and with the Mars Sample Return mission they will know how to get the samples back. Moreover, there is direct knowledge transfer of JPL's orbital mechanics and the trajectory tracking experience of the Deep Space Network that will be directly applicable to human flights.

Overall, the Mars 2020 mission is a transformative and transferrable scientific, engineering and operational experience for many reasons. And of course if the Perseverance rover finds evidence of ancient or extant life on Mars, that would forever alter our view of the universe and our place in it.

Appendix 1
Mars Exploration Program

This appendix will summarize the history and goals of the overall MEP program as it has evolved over the last three decades. This will put the current Mars 2020 Perseverance mission into historical context and demonstrate that, at least for the past 18 years, the ongoing mission is fulfilling the program's goals and objectives. The MEP can only be classified as very successful and satisfying to the scientific community worldwide.

A1.1 History

The Viking missions of the 1970s searched for answers to the "life question" by directly seeking evidence of biological activity in the upper 10 cm (4 in) of the Martian surface at two widely separated sites. In effect they were trying to hit a scientific "home run" on our first attempt at bat. The life detection experiments were arguably the best available, given early-1970s technology and our limited understanding of how to detect life in extreme environments. Unfortunately the results of the Viking in-situ life detection experiments were inconclusive. Given our then limited knowledge of Mars, the two Viking landing sites selected were reasonable starting places, but we now know that Mars is a remarkably diverse and dynamic planet, with many distinct regions that may differ significantly in their potential for harboring records of existing or ancient life. In retrospect, in spite of the bold approach pursued by the Viking project, Mars wasn't quick to yield its secrets and in the following years NASA and the scientific community have developed an improved framework for examining Mars (as has been done for Earth

M. von Ehrenfried, *Perseverance and the Mars 2020 Mission*,
Springer Praxis Books, https://doi.org/10.1007/978-3-030-92118-7

itself). What we have found is a planet exceedingly rich in landscape diversity, and what seems to be a preserved record of sediments that may be an indication of a major role of liquid water in its early history.

The next attempt to go to Mars was not until July 1988 when a pair of Russian spacecraft were sent to investigate Phobos, the larger of the two Martian moons. Phobos 1 was nominal until a planned communications session on September 2, 1988 failed to take place. Phobos 2 operated nominally throughout its cruise and Mars orbital insertion phases, gathering data on the Sun, interplanetary medium, Mars, and Phobos. Shortly before the final phase of the mission, during which it was to approach within 50 m (164 ft) of Phobos' surface and release two landers, one a mobile "hopper" and the other a stationary platform, contact was lost. The mission ended when the spacecraft signal failed to be successfully reacquired on March 27, 1989. The cause of the failure was determined to be a malfunction of the onboard computer.

The Mars Observer spacecraft, also known as the Mars Geoscience/Climatology Orbiter, was launched by NASA on September 25, 1992 with a payload to study the geology, geophysics, and climate of Mars from orbit. This was NASA's first Mars mission since the Vikings, 17 years earlier. To reduce costs, the spacecraft was based on a modified Earth-orbiting commercial communications satellite. All went will during the interplanetary cruise, but communication was lost on August 21, 1993, three days prior to orbital insertion. It appears a leak in the propulsion system, which was being pressurized for that maneuver, caused an explosion that disabled the spacecraft. Attempts to re-establish contact were unsuccessful. This loss prompted the formation of a cohesive new program.

In 1994 NASA announced the Mars Exploration Program (MEP), initially called the Mars Surveyor Program. It assigned the lead role for MEP implementation to JPL. The program is sponsored and funded by the NASA Headquarters' Science Mission Directorate. MEP is currently operating a number of rovers and orbiters on and around Mars. It also contributes to Mars missions conducted by national and international partners and is developing future missions. Scientific data and other information for all MEP missions are archived in NASA's Planetary Data System.

A1.2 First MEP phase 1994 to 2003

Mars Global Surveyor (MGS) was a NASA spacecraft launched in November 1996. It was a global mapping mission that examined the entire planet from the ionosphere down through the atmosphere to the surface. In the context of MEP, Mars Global Surveyor carried out atmospheric monitoring for sister orbiters as they performed aerobraking (during which it is essential to know how the upper atmosphere is behaving) and helped to identify potential landing sites for rovers and lander missions and relay the surface telemetry. MGS completed its primary mission in January 2001. Contact was lost on November 2, 2006 during its third extended phase. A faint signal was detected three days later indicating that it had entered "safe mode." After efforts to resolve the problem failed NASA officially ended the mission in January 2007.

Mars Pathfinder was launched in December 1996 and made the first successful landing on the planet since the Vikings. It comprised a lander and a lightweight robotic rover named Sojourner. As part of NASA's Discovery Program it was a proof-of-concept for advanced technologies such as airbag-mediated touchdown and automated obstacle avoidance, both later exploited by the Mars Exploration Rover missions. Mars Pathfinder was also remarkable for its extremely low cost relative to other robotic space missions to Mars.

The Mars Climate Orbiter (originally known as the Mars Surveyor '98 Orbiter) was a 638 kg (1,407 lb) probe launched in December 1998 in order to study the Martian climate, atmosphere, surface changes and to act as the communications relay for the companion Mars Polar Lander. However, on September 23, 1999 communication with the spacecraft was permanently lost as it attempted orbital insertion. It encountered Mars on a trajectory that took it too close to the planet and it was either destroyed in the atmosphere or entered an orbit around the Sun. An investigation attributed the failure to a measurement mismatch between two software systems, with NASA using metric units and spacecraft's manufacturer Lockheed Martin using imperial units.

Launched in January 1999, the 290 kg (639 lb) Mars Polar Lander was to study the soil and climate of a region near the south pole. On December 3, 1999, after the descent phase was expected to be complete, the vehicle failed to reestablish communication with Earth. A post-mortem analysis determined the most likely cause of the mishap was premature termination of the engine firing prior to the lander reaching the surface, causing it to crash at a high speed. The vehicle also carried two surface-penetrator probes for the New Millennium Program's Deep Space 2 mission, which also remained silent.

A1.3 The 2003 restructured program

The following are summarized and edited statements primarily from Dr. James B. Garvin, then Director for Mars Exploration and the Director for the Solar System Division at NASA Headquarters, and Orlando Figueroa, then Director for Mars Exploration and the Director for the Solar System Division in the Office of Space Science at NASA Headquarters. Some of the comments come from 'The NASA Mars Exploration Program: Assessment of Mars Science and Mission

Priorities' issued in 2003. *Bear in mind that their comments must be seen in the context of the period.*

As a result of the past decade's less than successful experiences, NASA and the scientific community reviewed the Mars program. The newly restructured MEP became a fundamentally science-driven program that focused on understanding Mars as a dynamic "system" and ultimately addressed whether life is (or once was) a part of that system. It further embraces the challenges associated with the development of a predictive capability for the Martian climate, and how the role of water, axial obliquity variations, plus other factors, could have influenced the environmental history of the planet.

This section describes the scientific strategy of the restructured Mars effort. As it must continuously evolve in response to scientific discoveries and the changing requirements of the scientific drivers that are developed by the general scientific community to NASA it is important to acknowledge that the present strategy is a living one. The foundation of the present strategy is often referred to as "Follow the Water," and this serves to connect fundamental program goals pertaining to the planet's biological potential, climate, the evolution of the solid planet, plus preparations for eventual human exploration. Achieving a balance is seen to be important within the present MEP, given the challenges associated with making an assessment of the biological potential and climate history of a distant object such as Mars.

On the basis of the knowledge revealed by the Mariner, Viking and Mars Global Surveyor missions, scientists know Mars (like Earth) has experienced dynamic interactions between its atmosphere, surface, and interior that are at least in part related to water. As NASA embarks on an intensive program of exploration of Mars, investigating the pathways and cycles of water has emerged as a strategy that may lead to a possibly preserved ancient record of biological processes, as well as to an understanding of the character of paleo-environments on Mars. In humanity's exploration of extreme environments here on Earth (the deep ocean, ice fields, geothermal sites, and so on), wherever there is liquid water below the boiling point, evidence of life has been identified. The presence of liquid water sometime and somewhere in the Martian past, coupled with other key variables (temperature, pressure, soil chemistry, atmospheric chemistry, etc.) makes it an attractive target in expanding the scientific understanding of life, its origins and diversity within the universe. In addition, other scientific drivers have emerged, including the use of Mars to provide absolute calibration of the timing of major solar system events.

One example of the difficulties associated with understanding a planet as complex as Mars is associated with the assessment of its biological potential. Searching for signs of existing or ancient life on Mars is fraught with seemingly insurmountable challenges. At 150 million km^2 (58 million mi^2) there is a tremendous surface area within which to search; roughly equivalent to the continental land mass of Earth. Furthermore, even after five major missions, comparatively little is known about the characteristics of the upper surface layer and of the impact of ultraviolet and cosmic radiation upon the surface environment. Thus, evidence of potential life may lay tens (or even hundreds) of meters within the naturally shielded shallow subsurface. Our knowledge of the shallow subsurface remains purely inferential, yet predictive models indicate there could be reservoirs of liquid water and such environments may be compelling localities for in-situ exploration.

A1.3.1 The systematic approach

The NASA scientific organizations reassessed their approach to the MEP in all the identified areas. It was almost a systems engineering approach to a complex planetary system. For example the search for life or life-generating environments on Mars requires a systematic approach through which the scientists can begin to understand the complex systems of geology, climate, and biological potential that constitute the "Mars System." In order to understand Mars as a dynamic system, one must first establish a global context of information about the planet, and then validate and expand that knowledge using increasingly narrowly focused surface investigations, ground-truthing, and targeted reconnaissance. Having achieved a strong foundation of orbital and surface reconnaissance and directed studies, one is able to make a well-informed selection of the most-promising local sites from which to obtain samples for return to Earth for comprehensive analysis. The MEP referred to this approach as "Seek, In-situ, Sample," using increasingly narrowed cycles of "seeking," first from orbit, then on the surface, followed by collection of well-selected samples for investigation in terrestrial laboratories.

This approach parallels the one used in exploration for minerals and other natural resources here on Earth. Petrochemical companies use satellite imagery of Earth to identify regions with chemical indicators of processes that concentrate valuable materials in certain geological settings. They follow up with localized analysis of those regions, before sending in the "wildcat" crews to drill for resources beneath the surface. In the case of Mars we are prospecting for water, signatures of life, or ancient environments conducive to life as we currently understand it.

In attempting to understand the "real" Mars, it is first necessary to inventory the key constituents of the Mars System. This step will determine the foundation, or context within which particularly difficult questions (most notably whether there was ever life on Mars) can be addressed.

Thanks to Viking and the ongoing Mars Global Surveyor mission, we now know Mars has experienced wide swings in its climate, potentially extensive periods in its past when liquid water was persistent at the surface in localized depressions, and that it might harbor an active subsurface hydrological system even today. In spite of the failures of the Viking surface laboratories to detect any signs of life, we can refine the approach and continue the quest with real prospects of making major strides during the current decade.

A1.3.2 Global mapping-the foundation for context

How will we attack the mysteries of Mars to determine whether it ever harbored life or ever experienced climate oscillations which mimic those of Earth? These questions have puzzled planetary scientists for decades and today, thanks to the mapping by the Mars Global Surveyor mission a refined and robust strategy has emerged.

For example, in order to aggressively address whether life ever existed within the Mars System, it is first necessary to have a systematically increasing database of knowledge about the global surface, atmosphere, hydrosphere (although it may be entirely frozen into a cryosphere), and interior. A comprehensive global inventory of the Martian surface (and shallow interior) then facilitates detection of localized "anomalies," places that are different in terms of chemistry, temperature, venting of important gases, or other factors. This "prospecting" step is a key initial part of our strategy. We must seek the most promising places to continue intensive local reconnaissance in the context of a global picture of Mars.

Thus, the first step in the MEP exploration strategy is to acquire adequate global reconnaissance using orbital remote sensing tools that not only define Mars in a global context but also tell us where to look in our refined search for "hot spots," places where the action of liquid water and possibly temperature have provided "fingerprints" that we can locate from orbit. The MGS mission that is currently mapping Mars in combination with the Mars Odyssey orbiter constitutes the first wave of systematic orbital reconnaissance. The aim is to continuously refine our search for areas on Mars where higher-resolution and landed investigations can best continue the exploration for those materials which offer the most promising prospects for resolving issues related to life, the timing of events, and climate history.

At issue is what percentage of Mars today might be identifiable as "hot spots." Recent findings from the imaging systems and spectrometers operated by Mars Global Surveyor suggest that there may be a few hundred to perhaps a thousand "hot spots" worthy of near-term intensive investigation at the surface. Isolating the handful of the most scientifically compelling instances poses a challenge to the reconnaissance elements of our unfolding program.

Data from Viking and MGS suggest that certain landscapes on Mars were likely sculpted by the action of liquid water. However, the same formations might also have been formed as a consequence of exotic processes associated with wind, or even explosive volcanism. We cannot discriminate between the possibilities until we can go to those regions and study them at "sample scales" both at the scale for definitive process identification and at microscopic scales for provenance studies. So the next step in refining our global understanding of Mars from this first wave of orbital reconnaissance is to conduct surface investigations at some of the most scientifically compelling sites. The surface-based investigations will validate and calibrate our global remote sensing data, and demonstrate that the mapping from orbit matches the reality of the surface. Data from MGS and Odyssey will reveal hundreds to thousands of the most promising regions for these intensive surface investigations.

A1.3.3 Surface investigation and ground-truthing

Due for launch in 2003 the Mars Exploration Rovers will take those next steps in making the discoveries that could lead in the future to determining whether or not life ever arose on Mars. The goal of the mission is to seek conclusive evidence of water-affected materials on the surface. Its two rovers are designed to effectively serve as robotic field geologists and they will provide the first microscopic study of rocks and soils on Mars. The Mars Pathfinder Sojourner rover analyzed eight rocks and soil patches with one inadequately calibrated instrument (APXS) in 83 days of surface activity. The twin MERs will study dozens of rocks with at least three different calibrated instruments as well as capturing spectacular contextual images together with mineralogy (from hyperspectral middle-infrared imaging). They will also have the mobility to wander up to 1 kilometer across the Martian landscape, measuring the chemical composition of the soils, rocks, and even the previously inaccessible interiors of rocks where unaltered material may lurk. Just as human field geologists study Earth by using a hammer to break open rocks, the MERs will employ a rock abrasion tool to scratch beneath the outer covering of rocks and look inside with microscopic resolution. Evidence from the meteorites from Mars recovered on Earth indicates that carbonates existed there in the past, at least at microscopic scales. What we require to find out is whether carbonates existed at or near the surface, or whether they were produced in association with biological processes. The MER robotic geologists will help answer this question. Finally, the MER perspective will allow for quantified calibration and validation of orbital remote sensing data at the surface that will hopefully yield the capacity to extrapolate to other places on the planet that seem similar to the MER landing sites.

By studying the rocks and soils in a "hot spot" region chosen from the MGS and Odyssey reconnaissance data, the MERs will tell us whether what we are seeing from orbit is what we anticipate and if not, what it may represent instead at least chemically. The MERs will link the surface chemistry and mineralogy with that surmised from orbit, and facilitate extrapolation across many different places on Mars.

Armed with this new knowledge and characterization of water-related geological regions on Mars, we will be able to climb to the next level of reconnaissance and in-situ investigations, including the search for life-bearing environments in those water-related regions as well as the surface record of climate.

A1.3.4 Targeted reconnaissance and landing site characterization

As the ultimate reconnaissance tool in the Seek, In-Situ, and Sample strategy the 2005 Mars Reconnaissance Orbiter will focus on the "hot spots" identified from MGS and Odyssey data. It will use new observational tools, some able to resolve beach-ball-sized objects, to search the landscape of telltale layers and materials associated with the action of liquid water for the most compelling indicators of climate variability and environments that would have been suitable for bearing life (warm, wet, chemically benign, etc.).

Recent evidence suggests that water-related mineral indicators may be detectable from orbit at certain specific infrared wavelengths, provided high enough spatial resolution is achieved. While debate lingers within the science community about resolution thresholds, imaging thousands of promising sites at ~30-cm resolution would allow discrimination of water-related sedimentation from that associated with explosive volcanism (layers of cemented ash), wind, or global dust settling. Thus, MRO will seek to produce a globally distributed set of panchromatic and hyperspectral images that isolate the dozen or so most compelling sites for an intensive surface-based exploration and sample return. When coupled with the first in-situ examination of two different potentially water-related sites on Mars provided by our twin MERs, our approach will allow us to build confidence we can predict how Mars operates under certain conditions and to demonstrate the past action of water in the otherwise hyper-arid desert of modern Mars. MRO is also to help us to develop the first understanding of modern water as it behaves within the present Martian atmosphere and how climatology operates on annual basis. In addition, MRO will try to characterize the shallow subsurface in search of water-related layers or deposits and other possible stratigraphic indicators of ancient water-related environments.

Although MGS and Odyssey may identify hundreds to thousands of interesting places on Mars only two can be visited by the twin MER rovers, in part because of their landing precision (50 km at best). MRO is designed to evaluate the most compelling places identified previously and to test their prospects at new scales, wavelengths, and using tools which measure the vertical structure of the shallow subsurface. The "hottest" places for intensive surface exploration identified from MGS and Odyssey will be exhaustively targeted by MRO so that, by 2006, a set of compelling localities will be established within a global scientific framework. Of these, the top two or three, in terms of their potential as biomarker sites, will serve as the bridge into the second phase of our strategy. Of course, compelling sites might include those localities where the climate record is best exposed in a physical and chemical sense.

A1.3.5 In-situ analysis and returning samples to Earth

The final part of Phase One of the new Mars Exploration Program will start after MRO has "fingered" where to go. If all goes as planned, in 2007 we will launch a precision-landed mobile surface laboratory, the Mars Science Laboratory (MSL), to the most promising of the targeted sites to operate for at least half a year. The planned development of a new suite of miniature analytical instruments for this mobile laboratory, which are tuned to questions of geochemistry and biological processes, will measure aspects of the surface and subsurface materials that are potentially linked with ancient life and the paleo-climate. These plans include a laser Raman spectrometer to focus our surface search for carbonates, a micron-resolution optical microscope to assess patterns of micro-scale features, and an instrumented drill that can penetrate to a depth of 2–3 m (6.5–10 ft) in search of buried ice or other "shielded" substances. It could also probe the subsurface to a depth of 100 m (328 ft) using ground penetrating radar or other electromagnetic sounding approaches. With cooperation from Mars, it could confirm the surface presence of water-related minerals and carbonates and their formation histories. And MSL will also benefit from the continuing refinement of how orbital remote sensing can be used as a pathfinder to the surface localities that offer the highest probability of harboring Martian "fossils" or other indicators of past life. It will serve as both a scientific and a technological pathfinder for the robotic sample return campaign that forms the ultimate step in our Mars Exploration Program.

This phase of in-situ analysis will incorporate technological advances that permit mobile surface laboratories to be landed within a few kilometers (or less) of any particular spot on Mars. By precision landing near to a telling site and having the longevity and mobility to explore as if we were there ourselves, we shall be able to extend our search for life and other scientific indicators to horizontal scales on the surface that will be measured in multiple kilometers rather than football fields. MSL will serve as the bridge to the next phase of Mars exploration: a future series of missions that will endeavor to bring preserved samples of the most interesting materials back to Earth, in context and with real prospects of their harboring bio-signs or chemical indicators of warmer and wetter past environments. It is clear a campaign of sample returns will be needed in order to tie the major Mars System events to the absolute chronology of the solar system. When Mars may have been more biologically hospitable and what the global planetary state of evolution was at that time are vital elements to achieving an understanding.

Thus, our refined strategy seeks to establish a suite of the most promising places for intensive surface analysis prior to the technological leap of returning samples to Earth for analysis. Once we have identified the hottest prospects, a program of long-duration and reasonably long-range mobile surface laboratories must unravel what is in the rocks, soils, ices, and atmospheric constituents that could be linked to favorable environments for biology and for its preservation, or indeed for other important questions. Under this strategy, with good fortune, by the close of 2008 we could be receiving images from Mars of microscopic features not unlike those identified by David McKay and Everett Gibson from the ALH84001 Allan Hills meteorite.

A1.3.6 Use of Discovery class missions

The overarching science thrust of the Mars Exploration Program is to examine the diversity of Mars by investigating multiple sites with mobile surface laboratories. Although it would need additional budget resources, a more aggressive program would allow the development of more than one mobile surface lab and enable the exploration of multiple "hot spots." It would also facilitate the earliest possible sample return missions. One way for MEP to achieve balance, innovation, and adaptability is the Mars Scout Program. Currently planned to start in 2007, this will solicit principal-investigator-led missions to explore Mars in focused ways that are not currently baselined in the core MEP program of flight missions. The aim is to promote scientific innovation within the MEP by allowing the broader community to compete for a "Discovery class" mission every other Mars launch opportunity. Given the somewhat limited breadth of the core MEP program and the recommendations of the Mars Exploration Payload Assessment Group, there is plenty of room for additional, high-science-value missions within the overall Mars Exploration Program. A 2007 Scout competition is expected to engage the scientific and engineering community and deliver an innovative mission with a specific focus not emphasized or treated within the core program. It is possible that the first Mars Scout mission will involve an array of small surface stations to explore the surface diversity of Mars, or an orbiter to map aspects of it globally that cannot be implemented on the 2005 MRO mission. A preliminary set of ten Mars Scout mission concepts were selected for 6 month study starting in June of 2001.

Finally, the first of several "informed" MSR missions is planned for a late 2011 launch, aiming to return ~1 kg of samples by 2014. The specific scientific scope of the first of these mission remains in the hands of science definition teams but the intent is to build upon the technologies utilized in 2007 for the smart mobile surface laboratory in order to selectively screen samples at a surface site selected to provide the best sedimentary record of water-related materials indicative of a hospitable paleo-environment. Given the material diversity of the planet and the challenges presented by sampling one scientifically compelling locality seeking definitive answers to the driving scientific questions about Mars, it is unlikely a single MSR mission to a sedimentary site will fulfill the scientific requirements and needs. Thus NASA plans to implement a campaign of MSR missions in the next decade (2011–2020) that will involve long-lived surface exploration having subsurface access in order to provide the most diagnostic materials for analytical investigations in terrestrial laboratories.

A1.3.7 Summary of the restructured program in 2003

Finding the right places on Mars to undertake the search for origins of life, the record of climate, and ultimately our place as humans within the cosmos is the first step required in the new Mars exploration strategy. A natural extension of this progressive strategy of orbital, surface, and ultimately return-sample-based reconnaissance is to visit two or more sites with precision-landed mobile labs by the end of the decade. Such a strategy, combined with small, totally competed missions in 2007 and beyond will extend our ability to seek elusive clues to the possibilities of life or at least for evidence of ancient, warm, wet environments. Ultimately the surface-based search, in the context of orbital "foundation" data sets, will yield Mars' secrets and allow us to return samples safely to Earth for unprecedented analytical scrutiny.

The new (2003) Mars Exploration Program will deliver a continuously refined view of Mars with the excitement of discovery at each step. What might we find as we move along the roadmap of mission events this decade? MGS has already identified possible signs of a source (albeit small) of liquid water at the surface. Odyssey could discover carbonates at the surface, or regions of enrichment in hydrogen, as well as evidence of possible artesian geothermal vents. The twin MERs may discover local evidence for how water once persisted at the surface and what ultimately we should seek from orbit. MRO may find ancient "oases," places where chemical and morphological evidence of past warmer and wetter environments is preserved. Alternately, it may find voluminous repositories of buried water ice in the shallow subsurface.

In 2008 the first of potentially several mobile surface laboratories could reveal materials that are indicative of locally warm, wet paleo-environments and make the first in-situ detection of organics. A 2007 Mars Scout could sample Martian atmospheric dust or probe the workings of Martian meteorology. Ultimately the first samples of rocks, soils, dust, and perhaps volatiles, will arrive on Earth by 2014. Analysis of these samples will "open the door" for human missions to the Red Planet.

NASA has fashioned a strategy that is risk attentive, with a natural responsivity to the science challenges that will emerge as discoveries are made. It is linked to our experiences exploring the deep ocean here on Earth, as well as a strategy that uses Mars as a natural laboratory for understanding life and climate on other planets.

A1.4 Mars mission successes

The fifteen year period from 2003 to 2018 saw success after success, but MSR is still a decade away. The following is a brief summary of the missions during this period.

The NASA Mars Exploration Rover (MER) mission began with the 2003 launch of the rovers Spirit and Opportunity. They landed at widely separate locations in January 2004 seeking geological evidence of past water activity. Both rovers far outlived their planned missions of 90 sols. MER-A Spirit was active until March 22, 2010 exploring Gusev Crater. MER-B Opportunity was active until June 10, 2018 driving a record 45 km (28 mi) across the dune fields of Meridiani Planum.

Mars Reconnaissance Orbiter was launched in August 2005 and reached Mars in March 2006. In November 2006, after five months of aerobraking, it achieved its final science orbit and began its primary program of reconnaissance for potential landing sites. When missions landed, it relayed their data back to Earth. It is still operating today, far beyond its intended design life. Owing to its critical role as a high-speed data-relay for ground missions, NASA intends to continue its service as long as possible. It has returned over 400 terabits of data and is supporting the Perseverance mission.

While not a MEP mission, the Phoenix Mars Lander was part of the Mars Scout Program and chalked up another success by landing in the polar region (without snow cover) on May 25, 2008. It operated for 157 sols until November 2nd. Its instruments were used to assess the local habitability and to research the history of water on the planet. The multi-agency program had international and industry partners and was led by the Lunar and Planetary Laboratory at the University of Arizona with project management by JPL. It was the first-ever NASA mission to Mars that was led by a public university. As available solar power dropped with the onset of Martian winter the lander eventually ran out of power. The mission was declared finished on November 10, 2008, after engineers were unable to re-establish contact. The lander was considered a success because it achieved all of the planned science experiments and observations.

The Mars Science Laboratory (MSL) was launched in November 2011 and on August 6, 2012 the new sky crane system gently lowered the Curiosity rover to the surface in Gale Crater. The science focuses on the habitability of the planet through time and collecting data to help prepare for a human mission. It is still providing useful data to the MEP after nearly a decade. (The way in which the MSL mission helped the design of the Mars 2020 mission and its Perseverance rover is described in Chapter 3.)

The Mars Atmosphere and Volatile EvolutioN (MAVEN) is an orbiter built by NASA to investigate the upper atmosphere and ionosphere of Mars to find out how the solar wind strips away volatile compounds. This research gives insight into how the planet's climate has changed over time. MAVEN was launched in November 2013 and entered orbit around Mars on September 22, 2014. It serves as a backup relay orbiter. It is continuing its science mission with all instruments still operating and with enough fuel to last at least until 2030.

Another of NASA's Discovery Program missions was the Interior Exploration using Seismic Investigations, Geodesy and Heat Transport (InSight). This is a lander to study the deep interior of Mars. Most of its science instruments were supplied by Europe. It was launched in May 2018 and landed on November 26, 2018. It was able to deploy a package on the surface to measure seismic activity and provide accurate 3-D models of the planet's interior, but a "mole" that was designed to "hammer" itself 5 m (16 ft) into the ground to measure internal heat flow in order to study the planet's early evolution proved unable to achieve any significant depth, largely because the properties of the soil were so dissimilar to those expected. InSight is approved for extended operations through December 2022.

A1.5 The present

The current program as stated on NASA's MEP website (as of October 2021) is summarized as follows:

Mission statement

The goal of the Program is to explore Mars and to provide a continuous flow of scientific information and discovery through a carefully chosen series of robotic orbiters, landers and mobile labs interconnected by a high-bandwidth Mars/Earth communications network.

About the program

NASA's Mars Exploration Program is a science-driven, technology-enabled study of Mars as a planetary system in order to understand:

- The formation and early evolution of Mars as a planet.
- The history of geological and climate processes that have shaped the planet through time.
- The potential for Mars to have hosted life (its "biological potential").
- The future exploration of Mars by humans.
- How Mars compares to and contrasts with Earth.

The science goals (as discussed in Chapter 2) are:

- Determine whether Mars ever supported life (and still does).
- Understand the processes and history of climate on the planet.
- Understand the origin and evolution of Mars as a geological system.
- Prepare for human exploration.

The programmatic goals of MEP are directly responsive to NASA's Strategic Plan issued in 2014, and include:

- Maintaining a continuous scientific presence at Mars.
- Providing continuing improvements in technical capabilities of robotic Mars missions.
- Capitalizing on opportunities that will contribute to the advancement of the knowledge that is required for future human exploration of Mars, in collaboration with NASA's Human Exploration and Operation Mission Directorate (HEOMD) and the Space Technology Mission Directorate (STMD).
- Ensuring that scientific measurements that can enable human exploration of Mars are considered for flight and that opportunities to fly instruments-of-opportunity and technology demonstrations from HEOMD and STMD are exercised on a mutually agreed upon basis.
- Supporting communications activities required for the successful conduct of MEP's core science mission and NASA's goals for helping to develop scientific literacy in the nation.

MEP derives its scientific goals from interactions with the planetary and Mars science community (e.g. through the Mars Exploration Program Analysis Group, or MEPAG). It has an evolving science strategy, with related MEP science goals that are consistent with the priorities in the 2011 Solar System Decadal Survey Vision and Voyages for Planetary Science in the Decade 2013–2022, conducted by the National Academy of Sciences' National Research Council. These goals are formed into specific requirements and, as appropriate, applied to individual missions in program-level project requirements. (See the References section of this book)

In order to support an integrated program structure, MEP undertakes activities that provide crosscutting functions and long-term investments for the future as follows:

- The Mars Data and Analysis Program (MDAP) is a scientific research and analysis effort that sponsors detailed studies of data returned from Mars to shape the next steps of future-mission studies. This research is intended to improve upon open science questions relating to current hypotheses. Brief synopses of such research can be found on the "NSPIRES" website using "MDAP" as the search keyword.

- The Program Formulation Office undertakes advanced studies relating to future missions. MEP sets priorities for these studies through informal (e.g. MEPAG) and formal channels (e.g. the National Academies' Solar System Decadal Surveys). Studies and planning for missions, as well as identification of technology needs and investments to meet them are the focus of multiple long-range program planning and targeted technology investments. Advanced studies also address possible needs to replenish Mars communication capabilities and other programmatic support (e.g. landing site certification) that may be needed to support future missions.

Orbiters with required capabilities and/or other priorities (such as sample return flight elements) may also provide opportunities to address scientific objectives. Missions that address other NASA objectives (e.g. to prepare for future human exploration) are considered jointly with key stakeholders from other parts of the agency. Another responsibility for the Program Formulation Office is definition of interfaces between ongoing missions or missions in development, and future missions which are not yet in development. This activity is especially important where Program objectives require coordination across multiple projects (e.g. the case of multiple missions cooperating on the return of samples from Mars).

- The MEP supports the implementation of both large and small directed missions as well as those that are competitively selected and led by a PI. These have included the Phoenix mission that landed in the north polar region of Mars to study ice and the MAVEN orbiter which is measuring the rate at which atmospheric gases are escaping from the planet and the implications for its ancient climate.
- Other crosscutting activities include optimizing overall science strategy, program risk management, and risk communication, telecommunication strategy, advanced capability development (i.e. technology and program infrastructure), interfaces with future missions, planetary protection, and communications with the target NASA audiences identified in the 2013 NASA Communications Framework.

So that brings the reader up to the present with the Mars Exploration Program.

As of late 2021 the Perseverance rover has started to collect samples which are currently expected to be retrieved by the first Mars Sample Return mission and delivered to Earth around 2031. Then scientists will finally be able to use their best investigative tools to make the ultimate determination about life on Mars, past and/or present. That work may go on for decades just like the research on lunar samples that were brought back a half century ago is still underway now.

Appendix 2
Conclusions By The Mars 2020 Science Definition Team

Background

This is the effort that officially kicked off what the science community wanted in what would be the Mars 2020 Perseverance mission. One can see from the actual mission now underway that it is well on its way to fulfilling the scientists' vision, goals and requirements. This appendix focuses on the points in the conclusion of the final report.

On July 9, 2013 NASA released a 154-page report that recommended sending a near-clone of the $2.5 billion Curiosity rover to Mars at the end of the decade in order to seek signs of past life, collect samples for eventual return to Earth, and demonstrate technologies that would assist future robotic and human exploration of the Red Planet.

NASA had chartered the Mars 2020 Science Definition Team (SDT) the previous January to scope out instruments for the proposed rover mission that would meet the project's budget, deadline and goals. NASA was confident it would be able to leverage off the Curiosity design to build a differently outfitted clone for about $1 billion less than the original. (That didn't turn out to be the case.)

The report states: "The SDT's evaluation of the 2020 opportunity for Mars finds that pioneering Mars science can be accomplished within the available resources and that the mission concept of a science caching rover, if implemented, would address the highest priority, community-vetted goals and objectives for Mars exploration. It would achieve high-quality science through the proposed suite of nested, coordinated measurements and would result in NASA's first Mars mission configured to cache samples for possible return to Earth at a later date."

John Grunsfeld, then NASA's Associate Administrator for science, confirmed the mission objectives determined by NASA with input from the Mars 2020 Science Definition Team "will become the basis, later this year, for soliciting

M. von Ehrenfried, *Perseverance and the Mars 2020 Mission*,
Springer Praxis Books, https://doi.org/10.1007/978-3-030-92118-7

proposals to provide instruments to be part of the science payload on this exciting step in Mars exploration."

The Mars 2020 Science Definition Team was led by Brown University professor Jack Mustard and was composed of 19 scientists and engineers from universities and research organizations, some of whom are currently supporting Perseverance.

Conclusions

The Mars 2020 rover, as envisioned by SDT, was to be the bold next chapter in over two decades of systematic exploration of a planet that may hold a record of past life. Numerous places on the surface have been found by orbiters to record past potentially habitable environments in their rock records. In-situ exploration of an extremely small sample of these environments by the two Mars Exploration Rover (MER) rovers and the Curiosity rover confirm past water and potentially habitable conditions. The scientific community has recognized that the next level of exploration of Mars' geological evolution, past habitability, and the search for signatures of past life requires more sophisticated laboratory measurements at a level achievable only in state-of-the-art laboratories on Earth. The Committee on the Planetary Science Decadal Survey (NRC, 2011) recommended that NASA's highest priority for large missions should be one that explores a key site on Mars and assembles a cache of samples for return for detailed analyses on Earth.

The Mars 2020 SDT has investigated whether a mission could take meaningful steps toward this grand objective within a constrained cost, while also providing more immediate results by characterizing in-situ the past habitability of one site and evidence for preservation of biosignatures. The SDT has also considered the potential for the mission to pave the way for future human exploration. It found that these objectives have such a high degree of overlap that they would be most efficiently addressed on a single mission.

For Objective A, explore an astrobiologically relevant ancient environment to decipher its geological processes and history, including the assessment of past habitability. The reasoning progressed as follows:

1. Deciphering and documenting the geology of a location requires in-situ, geological measurements and results from analyzing those measurements.
2. Rover imaging and compositional measurements should be of sufficient coverage, scale, and fidelity to permit their placement into the context of the orbital observations that provide the broader spatial coverage needed to understand regional geology.
3. A key strategy for interpreting past habitability is to seek geochemical or geological proxies for past environmental conditions as they are recorded in the chemistry, mineralogy, texture, and morphology of rocks.

4. Some aspects of the geological record of past habitable conditions may be neither preserved nor detectable. Therefore, inability to detect geological evidence for all four habitability factors (raw materials, energy, water, and favorable environmental conditions) should not preclude interpretation of a site as having been a habitable environment.
5. Five measurement types will be threshold (i.e. minimum) requirements to effectively and efficiently characterize the geology of a site and assess its past habitability: context imaging, context mineralogy, fine-scale imaging, fine-scale mineralogy, and fine-scale elemental chemistry.

For Objective B, assess the potential for biosignature preservation in the chosen geological environment and search for biosignatures. The following reasoning led to the SDT report's finding that a returnable cache of scientifically identified and selected materials was required in order to accomplish the science objectives:

1. Confidence in interpreting the origin(s) of potential biosignatures rises with the number of potential biosignatures identified, and with a better understanding of the attributes and context of each potential signature.
2. To thoroughly characterize and make a definitive discovery of Martian biosignatures would require the return of samples to Earth for analysis.
3. To study the potential for multiple types of biosignatures that might be preserved in multiple geological units which represent both a variety of potential past habitable environments and a range of potential states of preservation, at least four or five sample sets must be returned to Earth.
4. In-situ detection of organics would not be required for returning samples to Earth. Other valuable attributes could qualify samples for return, such as the presence of other categories of potential biosignatures or evidence of high preservation potential or a past habitable environment.
5. Six measurement types are threshold requirements to assess biosignature preservation potential and to search for potential biosignatures: context imaging, context mineralogy, fine scale imaging, fine-scale mineralogy, fine scale elemental chemistry, and organic matter detection. Notice that the first five threshold measurements are identical with those supporting Objective A and that in this case organic matter detection was added.

For objective C, demonstrate significant technical progress towards the return to Earth of scientifically selected and well-documented samples. This produced the following logical progression:

1. The SDT concurred with the detailed technical and scientific arguments made by the Decadal Survey (NRC, 2011) and MEPAG (most recently summarized in E2E-iSAG, 2011) for the critical role returned samples would play in the scientific exploration of Mars.

2. Significant technical progress by Mars 2020 towards the future return of samples to Earth required assembly of a cache of scientifically selected, well-documented samples packaged in such a way that it could be safely returned to Earth.
3. Although there are different ways of organizing the sample return steps (selection, caching, raising to orbit, and returning to Earth) into specific missions, the necessary first step in any scenario is always to select and cache samples.
4. Any progress toward this objective that doesn't create a returnable cache would have to be tackled by the next mission that makes progress toward sample return. Only through assembly of a returnable cache would that progress not need to be repeated by another mission. A returnable cache retires significant technical risk for sample return and thereby achieves a major milestone worthy of the efforts of spacefaring nations.
5. The SDT concluded that in order to achieve Objective C, the threshold science measurements would be those listed under Objective A, and the baseline measurements would include organic detection as baselined for Objective B.

For objective D, provide an opportunity for participation by HEOMD or Space Technology Program (STP), compatible with the science payload and within the mission's payload capacity. This inspired the following conclusions:

1. Three classes of environmental measurements are necessary to support HEOMD's long-term objectives: architecturally driven (in-situ resources, atmospheric measurements for EDL, etc.), safety driven (surface radiation, toxicity of materials, etc.) and operationally driven (surface hazards, dust properties, electrical properties, etc.).
2. The threshold and baseline measurements that address Objectives A, B, and C also each address a number of HEOMD strategic knowledge gaps.
3. Returned samples would address the objectives of HEOMD in relation to biohazards, the properties and toxicity of Martian dust, and the chemistry and mineralogy of regolith.
4. There are important opportunities for valuable technology development on Mars 2020 which would influence sampling, improved landing site access, planetary protection, improved science productivity, and risk reduction.
5. The CO_2 capture and dust characterization payload is HEOMD's expected contribution to the Mars 2020 mission. By adding weather measurements, that payload would also addresses synergistic science objectives.
6. The entry and descent phases of the 2020 mission should be characterized by a system with improvements over the MEDLI system used by the MSL mission.
7. The technologies associated with a Range Trigger should be a threshold capability and the SDT strongly encouraged inclusion of terrain-relative navigation as highest priority baseline capability to help ensure access to high priority sites and reduce science risk related to site selection.

These logical processes prompted the SDT to reach the following mission-level conclusions regarding the proposed Mars 2020 rover:

1. Significant technical progress by Mars 2020 towards the future return of samples to Earth requires assembly of a cache of scientifically selected and well-documented samples that are packaged in such a way that they can be safely returned to Earth.
2. To thoroughly characterize and make a definitive discovery of Martian biosignatures would require the return of samples to Earth for analysis.
3. Five core payload elements: two contextual measurements (imaging and mineralogy) and three types of contact measurements (fine-scale imaging, mineralogy, and elemental chemistry) jointly enable thorough analysis of whether the chosen site on Mars was once habitable.
4. Adding a sixth payload element. A search for preserved organic carbon would help to evaluate whether potential biosignatures of past life may exist.
5. These payload elements are the same as those to select and document the scientifically most important samples for caching. The SDT endorsed the strategy described by previous science panels for collecting and storing samples of key sedimentary, hydrothermal, and igneous rock materials.
6. The rover platform and the instruments on it addressed many gaps in the strategic knowledge required for future human exploration of Mars. The rover would also be a suitable platform for key technologies to improve landing accuracy, understand the local environment, and test methods to extract local resources which would further prepare not only for human exploration but also for the return of the sample cache to Earth.
7. The mission plan for completing these objectives in one Mars year was ambitious but would be feasible if the science instruments were efficient and the plan for exploring the site was chosen carefully.
8. Cost, the technology of caching, and the limitations of even the smartest robotic instrumentation together prevent creation of a single rover which could both cache as well as produce laboratory quality sample processing. Caching takes priority because that could lead ultimately to much greater scientific return using Earth-based laboratories.
9. Any Mars 2020 mission that does not provide a returnable cache would oblige a later mission to repeat key aspects in progressing toward sample return. Only the assembly of a returnable cache ensures that these tasks may not need to be repeated on another mission. Only a returnable cache also retires significant technical risk, and thereby would achieve a major milestone worthy of the efforts of spacefaring nations.

The SDT therefore concluded that the proposed Mars 2020 rover mission would be the best, most scientifically impactful next step in exploring the closest world at which humanity might answer the question of whether life arose elsewhere in the solar system.

Appendix 3
Mars 2020 Mission Team Biographies

The current Mars 2020 mission team comprises at least five hundred people, and there are probably over a thousand more having something to do with the mission worldwide. This appendix could not possibly include all of their biographies but it features those who are often quoted, appear in videos, or are in key positions. My apologies to the others but their contributions are duly noted and they should take great pride in their support for such a truly historic mission. The people are listed in alphabetical order within each category.

The Mars 2020 Perseverance team consists of scientists, engineers, technicians and support personnel from multiple disciplines with international participation from countries and organizations around the world. The science team includes principal investigators from the US, Spain, and Norway.

A3.1 NASA Headquarters

Dr. Lori Glaze is Director of the Planetary Science Division of NASA's Science Mission Directorate. It is focused on scientific research and space flight missions that address fundamental questions about the formation and evolution of the solar system, including understanding planetary environments that can (or could have in the past) support life.

Prior to NASA Headquarters, Dr. Glaze led the Planetary Geology, Geophysics and Geochemistry Laboratory at Goddard Space Flight Center in Maryland and was the Deputy Director of Goddard's Solar System Exploration Division.

M. von Ehrenfried, *Perseverance and the Mars 2020 Mission*,
Springer Praxis Books, https://doi.org/10.1007/978-3-030-92118-7

Fig. A3.1.1 The Mars 2020 Team. Photo courtesy of NASA/JPL-CalTech

Fig. A3.1.2 Dr. Lori Glaze

Her research interests include physical processes in terrestrial and planetary volcanology, atmospheric transport and diffusion processes, and geological mass movements. Her work focuses on data analysis and theoretical modeling of surface processes on all the terrestrial solar system bodies, particularly the Earth, Venus, Mars, Moon, and Io. Dr. Glaze was a member of the Inner Planets Panel during the most recent Planetary Science Decadal Survey and has had a role on the Executive Committee of NASA's Venus Exploration Analysis Group (VEXAG) for several years, serving as its Chair from 2013 to 2017. Dr. Glaze was a member of the Planetary Science Subcommittee from 2011 to 2013.

Dr. Glaze graduated from the University of Texas, Arlington with a B.A. and M.S. in Physics. She received a Ph.D. in Environmental Science from Lancaster University in England. She has also worked at JPL, and at Proxemy Research as Vice President and Senior Research Scientist.

Fig. A3.1.3 Jeffery Gramling

Mr. Gramling is Director of the Mars Sample Return Program within the Science Mission Directorate at NASA Headquarters. He is also currently working on the Mars 2020 mission with respect to the sampling effort. He began his career at the NASA Goddard Space Flight Center in 1985 developing hardware simulators for the Hubble Space Telescope Project and in 1992 became a subsystem manager on the Tracking and Data Relay Satellite (TDRS) Project. After working on several satellites he was appointed Project Manager for the TDRS-K, L, M development program, successfully launching TDRS-K in 2013 and TDRS-L in 2014.

In 2015 Mr. Gramling became Associate Director of Flight Projects for Earth Science research at Goddard, concurrently serving as Program Manager of the Earth Systematic Missions Program (ESMP). He joined the Applied Physics Laboratory at Johns Hopkins University in 2018 as the Project Manager of the Galactic/Extragalactic ULDB Spectroscopic Terahertz Observatory (GUSTO) during its Preliminary Design Review and Confirmation Review. Mr. Gramling holds a B.S. in Computer Engineering from Clemson University and an M.S. in Electrical Engineering from the University of Maryland. He is a Senior Member of the Institute of Electrical and Electronics Engineers, and Associate Fellow of the American Institute of Aeronautics and Astronautics.

Fig. A3.1.4 Eric Ianson

Mr. Ianson became the Mars Exploration Program Director in November 2020, while maintaining his position as Deputy Director, Planetary Science Division (PSD). He delivers strategic leadership in coordination with the PSD Director for all NASA activities pertaining to the exploration of the solar system, including science missions, research programs and technology development across several NASA Centers, international and interagency partners, academia, and industry.

From 2015 to 2019, Mr. Ianson was Associate Director for Flight Programs, Earth Science Division at Goddard Space Flight Center where he coordinated all aspects of the NASA Earth Science flight program. Prior to that he was Deputy Associate Director, Earth Science Projects Division, involved with all NASA Earth science flight project activities undertaken by Goddard. Since joining NASA in 2004, he has supported many different projects.

Mr. Ianson earned a B.S. in Mechanical Engineering from the University of Rochester in 1990 and an M.S. in Aeronautical Engineering from the University of Southern California in 1993.

Fig. A3.1.5 Dave Lavery

Mr. Lavery is the Program Executive for Solar System Exploration at NASA. As a roboticist, he also is a member of the Executive Advisory Board of FIRST (For Inspiration and Recognition of Science and Technology) and mentored Team 116 in the FIRST Robotics Competition. He headed NASA's Telerobotics Technology Development Program, responsible for the direction and oversight of robotics and planetary exploration within the organization. He has been heavily involved in all Mars rovers. He manages NASA's Robotics Alliance Project and is involved with the National Robotics Engineering Consortium.

Fig. A3.1.6 Dr. Michael Meyer

Dr. Meyer has been the Lead Scientist for NASA's Mars Exploration Program at NASA Headquarters. He oversees the program's science operations and planning. He was the Program Scientist for the 2001 Mars Odyssey mission and the Senior Scientist for Astrobiology at NASA Headquarters. As the Program Scientist he is responsible for developing mission science requirements, and then working with the mission scientists and engineers to maximize the science that is achieved.

Dr. Meyer was the Program Scientist for the Mars Microprobe mission. He was also the NASA Planetary Protection Officer responsible for mission compliance to NASA's policy concerning forward and back contamination during planetary exploration.

Before NASA, Meyer was an assistant research professor at the Desert Research Institute and an associate director and associate in research for the Polar Desert Research Center at Florida State University. Meyer has undertaken research in microorganisms which live in extreme environments such as the Gobi Desert of Mongolia, Siberia, and the Canadian Arctic. He is also a veteran of six research expeditions to Antarctica. His experience also includes two summers as a treasure salvager off the coasts of Florida and North Carolina. Dr. Meyer earned his M.S. and Ph.D. in Oceanography from Texas A&M University and his B.S. in Biology from Rensselaer Polytechnic Institute.

Fig. A3.1.7 Dr. Mitchell Schulte

Dr. Mitchell Schulte is a Program Scientist with the Mars Exploration Program (MEP) as well as the Planetary Science Division (PSD) in the Science Mission Directorate at NASA Headquarters. As a Program Scientist he is responsible for and manages the science content of NASA's missions to Mars. He is currently working on the Mars 2020 Perseverance mission. He has also overseen the US contribution to the Mars Organic Molecule Analyzer (MOMA) intended for the European Space Agency/Roscosmos ExoMars rover mission. He also leads Mars Research and Analysis for MEP, manages the Mars Data Analysis Program, and serves as a discipline scientist for other PSD programs, in particular Exobiology and Habitable Worlds.

As a researcher, Dr. Schulte has focused primarily on understanding the geology and geochemistry of hydrothermal environments and the microbiological life in them. He has extensive field experience in deep sea hydrothermal vents, Iceland, Yellowstone National Park, and ophiolite terrains in Northern California. He also is interested in biosignatures and life detection in ancient Earth and extraterrestrial samples, having field experience of ancient microfossil sites in Western Australia. He has an A.B. and a Ph.D., both in Earth & Planetary Sciences from Washington University in St. Louis, Missouri.

George Tahu is the Program Executive at NASA Headquarters for the Mars 2020 Perseverance mission and he is also serving as Deputy Director of NASA's Mars Exploration Program. In addition to exploring Mars, he has led the development of robotic lunar landers, astrobiology Cubesat payloads, and science instruments contributed by NASA to international partner missions.

Prior to joining NASA, Mr. Tahu was a space project manager for the US Army Space and Missile Defense Command and served as space policy specialist at the Department of Commerce. He also conducted space policy analysis and research with ANSER's Center for International Aerospace Cooperation.

Fig. A3.1.8 George Tahu

Fig. A3.1.9 Jim Watzin

Jim Watzin is Director of the Mars Exploration Program. He recently succeeded Jim Green, NASA's Chief of Planetary Sciences. He was technical director and deputy program executive for Command, Control, Communication, Computer, Intelligence, Surveillance and Reconnaissance at the Missile Defense Agency in Huntsville, Alabama.

Watzin graduated from the University of South Carolina in 1978 with a B.S. in Mechanical Engineering. In 1980 he got his M.S. in Aerospace Dynamics and Control from Purdue. He joined NASA's Goddard Space Flight Center in 1980.

Watzin has led multiple flight projects and program offices, serving as NASA's program manager for several programs which included Living with a Star, Solar Terrestrial Probes, and Robotic Lunar Exploration. He established the Planetary Projects Division at Goddard Space Flight Center, oversaw the development of the Sample Analysis at Mars payload for the Curiosity rover, and mentored the MAVEN and OSIRIS-REx mission formulation teams.

Fig. A3.1.10 Dr. Thomas Zurbuchen

Thomas Zurbuchen is the NASA Headquarters Deputy Associate Administrator for the Science Mission Directorate and as such has been very active in the Mars 2020 mission. Growing up in Switzerland, Dr. Zurbuchen was an eager observer of the natural world from an early age. His curiosity led him to pursue degrees in Physics. He has served on, and led, innovative scientific teams that have helped enlarge our perspective of the solar system and the universe. Previous points of focus have been the planet Mercury and our Sun. He was also a professor of Space Science and Aerospace Engineering at the University of Michigan in Ann Arbor. He founded its Center for Entrepreneurship at the College of Engineering and developed and ran several campus-wide innovation initiatives, one of which led to the top-ranked undergraduate entrepreneurship program nationally.

Dr. Zurbuchen has authored or co-authored more than 200 articles in solar and heliospheric phenomena. He earned his Ph.D. and M.S. degrees in Physics from the University of Bern in Switzerland. His honors include multiple NASA group achievement awards, induction as a member of the International Academy of Astronautics, a NASA Outstanding Leadership Medal, and the 2018 Heinrich-Greinacher prize, the latter being the leading science-related recognition by the University of Bern.

A3.2 JPL Perseverance Team

Individuals on other teams are also considered to be part of the Mars 2020 team, as listed in Sections A3.3 and A3.4.

Fig. A3.2.1 Raymond Baker

Raymond Baker has been with JPL for over 20 years. He is an Entry, Decent and Landing Systems Engineer and also Senior Propulsion System Engineer. He was the Mars 2020 Motor Control Subsystem Deputy Product Delivery Manager prior to working on Perseverance. His other positions included EDL Product Delivery Manager, Acting Propulsion Product Delivery Manager, Propulsion Engineer and AutoNav Verification and Validation Lead.

Mr. Baker earned a B.S. in Aerospace, Aeronautical and Astronautical/Space Engineering and an M.S. in Mechanical Engineering, both from CalTech.

Fig. A3.2.2 Dr. Tanja Bosak

Dr. Bosak is a member of NASA's Mars 2020 Project Science Group and was on the Perseverance sample caching committee to determine the science strategy for caching samples. She is the professor of Geobiology in the Department of Earth, Atmospheric and Planetary Sciences at MIT and the group leader of the Program in Geology, Geochemistry and Geobiology. Her work in experimental geobiology asks how microbial processes leave chemical, mineral, and morphological signals in sedimentary rocks. Her laboratory research combines microbiology, materials science and sedimentology to explore modern biogeochemical/sedimentological processes, interpret the co-evolution of life and the environment during the first 80% of Earth's history, and search for signs of past life or prebiotic processes on Mars.

Dr. Bosak earned a B.S. in Geophysics from Zagreb University and a Ph.D. in Geobiology from the California Institute of Technology. Bosak's awards include the 2007 Subaru Outstanding Woman in Science Award given by the Geological Society of America and the 2011 James B. Macelwane Medal from the American Geophysical Union.

Fig. A3.2.3 Dr. Robert D. Braun

Dr. Braun was the Mars Sample Return Program Manager from April 2020 to March 2021. He is currently the Director for Planetary Science at JPL. He has management responsibility for the portfolio of planetary science formulation, technology, implementation and operations activities and as such he provides support to the Mars 2020 program.

Dr. Braun has more than 30 years of experience as a space systems engineer, technologist, and organizational leader. He is an acknowledged authority in the development of entry, descent and landing systems, and has contributed to the formulation, development, and operation of multiple space flight missions. He previously served as Dean of the College of Engineering and Applied Science at the University of Colorado Boulder, a faculty member of the Georgia Institute of Technology, and a member of the technical staff of the NASA Langley Research Center. In 2010–2011, Dr. Braun served as the first NASA Chief Technologist at NASA Headquarters.

He holds a Ph.D. in Aeronautics and Astronautics from Stanford University and an M.S. in Astronautics from the George Washington University as well as a B.S. in Aerospace Engineering from the Pennsylvania State University.

Fig. A3.2.4 Allen Chen

Allen Chen is a systems engineer in JPL's Entry, Descent, and Landing Systems and Advanced Technologies group and was the EDL Lead for Mars 2020. During his ten year tour of duty on the Mars Science Laboratory mission he was the EDL Operation Lead, EDL Flight Dynamics Lead, co-led the joint science/engineering Mars atmosphere characterization team, served as a member of the Flight System Systems Engineering team, and did play-by-play commentary for landing. He also worked on the Mars Exploration Rovers project, performing EDL reconstruction analysis and testing.

He joined the Mars Science Laboratory Curiosity EDL team in 2002, then joined the Mars 2020 team in 2013 soon after that was assembled. As EDL Lead he was responsible for ensuring the spacecraft traveled safely from the top of the Martian atmosphere to landing in Jezero Crater. Chen narrated the landing procedures with Guidance and Controls Operations Lead Swati Mohan on February 18, 2021. His callout "Touchdown confirmed. We're safe on Mars!" prompted the JPL Control Center to erupt in celebration, with team members hugging, high fiving, clapping and crying.

Chen studied Aerospace and Systems Engineering at MIT, and in 2007 received a Fully Employed Master of Business Administration (FEMBA) from the Anderson School of Management at the University of California, Los Angeles.

Fig. A3.2.5 Dr. Kenneth Farley

Dr. Kenneth Farley is the Mars 2020 Project Scientist. As such, he worked with engineers to design the Perseverance rover. He is also the W.M. Keck Foundation Professor of Geochemistry in the Division of Geological and Planetary Sciences at the California Institute of Technology.

His research centers on development and application of geochemistry techniques, especially involving isotopes of the noble gases to a wide range of terrestrial and solar system questions. His specific interests include geochronology of both Earth and Mars, the geochemical evolution of the Earth, and the behavior of noble gases in minerals.

Dr. Farley began his professorial career at CalTech in 1993 and over the years he has received numerous awards and honors. In 2000 he was awarded the National Academy of Sciences' Award for Initiatives in Research for the innovative and potentially beneficial nature of his research. He was elected to National Academy of Sciences and the American Academy of Arts and Sciences. He earned his B.S. (summa cum laude) in Chemistry at Yale University in 1986 and his doctorate in geochemistry at the University of California, San Diego in 1991.

Fig. A3.2.6 Dr. John Grant

Dr. John A. Grant III joined the Smithsonian in the fall of 2000 as a geologist at the Center for Earth and Planetary Studies at the National Air and Space Museum. He was a member of the science teams for the Mars Exploration Rovers Spirit and Opportunity, the Mars Science Laboratory Curiosity, and the HiRISE camera on Mars Reconnaissance Orbiter. On Spirit and Opportunity he served as a Science Operations Working Group Chair responsible for leading the day-to-day science planning of the rovers, whereas on Curiosity he is a Long Term Planner focusing more on achieving strategic goals for the mission. He is also a co-investigator on the HiRISE camera and is the Science Theme Lead for Landscape Evolution and Future Landing Sites. He served as co-chair for the science community process in selecting the landing sites for the Spirit, Opportunity, and Curiosity rovers and he was a co-lead in the process for selecting the landing site for the Mars 2020 rover.

Dr. Grant attended the State University of New York College at Plattsburgh and received his B.S. degree (magna cum laude) in Geology in 1982 and went on to earn an M.S. and Ph.D. in Geology at the University of Rhode Island (1986) and Brown University (1990), respectively. His doctoral dissertation focused on the degradation of meteorite impact craters on Earth and Mars. He remains interested in understanding the processes responsible for shaping planetary landscapes. He was the 2017 recipient of the G.K. Gilbert award given by the Planetary Geology Division of the Geological Society of America for his outstanding contributions to the solution of fundamental problems in planetary geology.

Fig. A3.2.7 Louise Jandura

Louise Jandura is the Sampling and Caching Chief Engineer for the Mars 2020 Perseverance rover and has served in this capacity throughout the entire project lifecycle, starting in 2013 with the architecture phase, proceeding to the design and delivery of the Sampling & Caching Subsystem and is currently supporting the sampling operations. Louise served in a similar capacity for the sampling system on the Curiosity rover. She is a Principal Mechatronics Engineer at JPL with expertise in space mechanisms and sampling systems. During her 30 plus years at JPL she has also worked on many flight projects including the Shuttle Radar Topography Mission, Genesis, Mars Exploration Rovers, and Aquarius. She received her B.S. and M.S. degrees in Mechanical Engineering from MIT.

Fig. A3.2.8 Dr. Gerhard Kruizinga

Dr. Kruizinga was the Navigation Team Chief for Mars 2020, responsible for all navigation from launch to arrival at the top of the Mars atmosphere. He has also worked as a navigation engineer on the Earth science missions TOPEX/Poseidon, Jason-1, GRACE and GRACE-FO. Also included are missions to the Moon such as Chandrayaan-1 & 2, LCROSS and GRAIL and missions to Mars including the Mars Global Surveyor, Odyssey, Mars Exploration Rover, Phoenix, Mars Orbiter Mission, Mars Science Laboratory, InSight and Mars 2020 as well contributing to the New Horizons mission to Pluto.

He graduated from the Faculty of Aerospace Engineering at the Delft University of Technology in the Netherlands with an M.S. Degree and then gained his Ph.D. from the University of Texas.

Fig. A3.2.9 Kimberly Maxwell

Kimberly Maxwell joined JPL in 2010 to work on the Mars Science Laboratory, which was then in development. She wrote the Activity Dictionary for MSL and then transitioned into a number of operational roles over the years in addition to taking over the Sample Analysis on Mars (SAM) Instrument Engineer's position shortly after Curiosity began science operations. After a brief stint on the Europa Lander project, she joined the Mars 2020 Perseverance Science Operations team. As the Science Operations Team Lead since January 2020 she helped ensure the mission could achieve its science goals and that the science team members were prepared for mission operations once the rover landed.

She earned a B.S. in Aeronautics and Astronautics from MIT in 2001 and then an M.S. in Aerospace Engineering in 2005 and an M.S. in Systems Architecting and Engineering 2009, both from the University of Southern California.

Dr. McNamee was the JPL Mars 2020 Perseverance Project Manager from 2013 when the program was first announced until the successful landing and checkout of the rover on Mars in February 2021. He was the Manager of the Interplanetary Network Directorate from 2009 to 2013. From 2002 to 2009, Dr. McNamee held several management positions including the Deputy Manager, Mars Exploration Program and Deputy Manager, Solar System Exploration Directorate and served in an interim capacity as Deep Impact Project Manager, Dawn Project Manager, and Discovery/New Frontiers Program Manager.

In 1998, Dr. McNamee was appointed Project Manager for Outer Planets/Solar Probe which encompassed three planned missions: Europa Orbiter, Pluto-Kuiper Express and Solar Probe. In parallel he continued in his role as Project Manager of the 1998 Mars Surveyor mission, a position he had begun in May 1995. Prior to

Fig. A3.10 Dr. John B. McNamee

that he was manager of the Mars Exploration Pre-projects and Mission Design Manager for Mars Pathfinder. From 1989 to 1992 he was the Engineering Office Deputy Manager and Navigation Team Chief for the Magellan mission to Venus, work that earned him NASA's Exceptional Service Award. After the launches of the Mars '98 Climate Orbiter in 1998 and the Mars '98 Polar Lander in 1999 he assumed management of Outer Planets/Solar Probe on a full-time basis.

Dr. McNamee earned a B.S.B.A in Economics from the University of Florida in 1975, and went on to receive an M.S. and Ph.D. from the University of Texas in Austin.

Dr. Mohan emigrated from India to the United States when 1 year old, and was raised in the Northern Virginia and Washington DC metro area. She gained her B.S from Cornell University in Mechanical & Aerospace Engineering and then attended MIT for her M.S. and Ph.D., both in Aeronautics/Astronautics. She has worked on multiple missions such as Cassini (mission to Saturn) and GRAIL (a pair of formation-flown spacecraft to the Moon). She has worked on Mars 2020 since almost the beginning of the project in 2013. She is currently the JPL Mars 2020 Guidance, Navigation, and Controls Operations Lead.

Fig. A3.2.11 Dr. Swati Mohan

Fig. A3.2.12 Dr. Katie Stack Morgan

Dr. Morgan is a research scientist at JPL with interests in geological mapping of planetary surfaces. Her research focuses on the Martian sedimentary rock record, using orbiter and rover image data to understand the evolution of ancient surface processes on Mars. Morgan has been a member of the Mars Science Laboratory Science Team since 2012 and is currently a funded Participating Scientist on the mission. She has been the Deputy Project Scientist of the Perseverance mission since 2017.

Previously, Dr. Morgan was Research Scientist, Geophysics and Planetary Geosciences Group (2014-present) and Collaborator, Science Office, Mars Science Laboratory (2012–2016).

Dr. Morgan has a B.A. in Geosciences and Astronomy from Williams College in Williamstown, Massachusetts (2008) and an M.Sc. (2011) and Ph.D. in Geology (2015), both from CalTech.

Fig. A3.2.13 Dr. Matthew Robinson

Dr. Robinson is a senior member of the Robotic Systems Staff Group in JPL's Mobility and Robotic Systems Section, which he joined back in 2001. He has more than 20 years of experience of Mars surface missions. He was the Deputy Manager in charge of development and delivery of the Sampling and Caching System for the Mars 2020 Perseverance rover. After launch, he transitioned to Strategic Sampling Operations Lead Engineer responsible for ensuring that the

Sampling and Caching System was ready for surface activities. Previously, he was awarded NASA's Exceptional Achievement medal for his contributions to the Curiosity mission as the Lead Robotic Arm Systems Engineer and Robotic Arm Surface Operations Lead. He also contributed to the Mars Phoenix mission as the Lead Robotic Arm Flight Software Engineer, Robotic Arm Engineer and Surface Operator.

He holds a B.S (1996) and a Ph.D. (2001) in Mechanical Engineering from the University of Notre Dame, Indiana.

Fig. A3.2.14 Keith Rosette

Mr. Rosette is the JPL Manager of the Mars Sample Return/Earth Return Orbiter Components Office. It addressed the management, development, fabrication and delivery of the Sampling and Caching System of the Perseverance rover which is to collect and store cores of various rock types for eventual return to Earth on the Mars Sample Return mission.

His previous posts include Deputy Manager of JPL Flight Systems Engineering, Integration & Test Section and a Product Delivery Manager. From 2005 to 2009 he was a Senior Mechanical Engineer at JPL. He then worked at Orbital Sciences Corporation as a Spacecraft Lead Systems Engineer for five years. He has a B.S. and M.S in Mechanical Engineering from Virginia Tech.

Fig. A3.2.15 Jessica Samuels

Ms. Samuels is the Surface Systems and Mission Manager for the Perseverance rover. She is currently a Technical Group Supervisor for the System Verification and Validation Engineering Group in JPL's Flight System, Integration and Test Section. She has supported the Mars Exploration Rover, Deep Impact and Mars Science Laboratory missions in Integration and Test, Systems Engineering and Operations roles. She holds B.S. degrees in both Aeronautical Engineering and Mechanical Engineering from University of California, Davis.

Ms. Spanovich has worked at JPL since 2005. She is currently Curiosity Rover Science Operations Team Chief and hence is responsible for the processes and procedures followed by the Curiosity Science Team. As the Mars 2020 Science Systems Engineer she supports the development of the science requirements for the mission. Prior to her work on Curiosity and Mars 2020 she served in many roles on Mars Exploration Rover, such as Uplink Lead for various instruments, Atmospheric Science Lead and the Science Operations Support Team Chief. In her early career, she supported Spirit and Opportunity rover remote operations from Arizona and also worked with the Phoenix Lander's engineers to develop software to calibrate science cameras. She got her B.S. in Astronomy from the University of Arizona and an M.S. in Aeronautical Science from Embry-Riddle Aeronautical University.

Fig. A3.2.16 Nicole Spanovich

Fig. A3.2.17 Dr. Adam Steltzner

Dr. Steltzner is the Chief Engineer of the Mars 2020 mission. He worked on the "sky crane" landing system that was developed for the Curiosity rover and later used by Perseverance. He joined JPL in 1991, in the Spacecraft Structures and Dynamics group. He worked on several flight projects including the Shuttle-Mir Program, Galileo, Cassini, Mars Pathfinder, Mars Exploration Rover and Mars Science Laboratory. Initially employed as a structures and mechanics person he gravitated towards Mars Entry, Descent and Landing systems. He served as the landing systems engineer on the canceled comet mission Champollion and was the MER mission Mechanical Systems Lead for EDL.

Dr. Steltzner gained a B.S. in Mechanical Engineering-Mechanical Design from the University of California, Davis in 1990, an M.S. in Applied Mechanics from CalTech in 1991 and a Ph.D. in Engineering Mechanics from the University of Wisconsin-Madison in 1999.

Fig. A3.2.18 Dr. Vivian Z. Sun

Dr. Sun has served as a Science Operations Systems Engineer on the Mars 2020 Science Operations Development Team since 2017. She is currently involved in the Perseverance campaign for planning and implementation operations. She was involved in geological and mineralogical investigations of the Mars 2020 landing site candidates. Her work also involved science operations training, including the design and implementation of procedures and training materials and tools used by the science team. In her involvement with Mars Science Laboratory Curiosity she led the team that decided the observations and activities that would be required in order to address specific hypotheses about the Vera Rubin Ridge terrain. She was the liaison between Rover Planners and the science team.

She has a B.S. in Planetary Science from CalTech in 2012 and gained an M.S. in Geological Sciences in 2014 and a Ph.D. in Earth, Environmental and Planetary Sciences in 2017, both from Brown University.

Fig. A3.2.19 Jennifer H. Trosper

Ms. Trosper is the Project Manager for the Mars 2020 Perseverance rover. In the thirty years since her initial hiring at JPL as a power subsystem engineer she has held critical engineering leadership roles on every spacecraft ever to have roved the surface of Mars. She joined the Perseverance team in 2015 and held several leadership roles during the development and operations of the mission. Initially, as the Mission System Development Manager and Surface Phase Lead, she was responsible for the team to envision and design the rover's onboard autonomous capabilities and state-of-the-art ground system necessary to meet the challenging science objectives of the surface mission. She then transitioned to the role of the Project System Engineer and Engineering Technical Authority, in which she led the project-wide systems engineering teams, verification and validation program, and technical risk assessment for the Launch, Cruise, Entry Descent & Landing, and Surface Operations capabilities during the integration and testing phase of the mission's development. Shortly prior to launch she transitioned to Deputy Project Manager and continued her role in leading the surface development. Additionally, she supported the oversight of cruise operations and EDL development as well as assisting with the overall budget, schedule, and workforce management. Now that operations are in progress she leads the project to acquire a scientifically valuable and diverse cache of samples for a future return to Earth.

In her prior roles on JPL missions, Ms. Trosper was the Deputy Project Manager and Mission Manager for the Mars Science Laboratory Curiosity rover. She also provided leadership of systems engineering and operations for Mars Exploration Rover, SMAP, and Mars 2001 Odyssey. In addition, she served as a testbed and subsystem engineer on the Mars Pathfinder and Cassini missions. She originally joined JPL as a subsystem engineer in power, attitude control and command data handling. Coupled with her experience as a testbed engineer this formed the basis for her transition into systems engineering. Ms. Trosper's end-to-end expertise in leading engineering teams with complex systems through design, verification and validation, and operations has been the hallmark of her success at JPL. She is also a key leader in infusing autonomy and state-of-the-art ground operations systems into JPL rover missions.

Ms. Trosper holds a B.S. in Aerospace Engineering from MIT and an M.S. in the same from the University of Southern California.

Fig. A3.2.20 Dr. Vandi Verma

Dr. Verma is Chief Engineer for Robotic Operations for the Perseverance rover. As the Assistant Section Manager for Mobility and Robotics Systems at JPL she specializes in space robotics, autonomous robots and robotic operations. She has worked upon a number of space robotics and artificial intelligence research and technology development tasks and has designed, developed and operated rovers on Mars and in the Arctic, Antarctica and Atacama Desert.

She has been working at NASA since graduating with a Ph.D. in Robotics from Carnegie Mellon University in 2005. She works on new capabilities from early design, through development, testing, launch and landing to surface operations. Since 2008 she has been driving rovers on Mars including Spirit, Opportunity, Curiosity, and now Perseverance, and operating the robotic arm and sampling system. She has written flight software for Curiosity and for Perseverance and simulation software used in operations.

Fig. A3.2.21 Matthew T. Wallace

Mr. Wallace is currently JPL's Deputy Director for Planetary Science, having served as the Mars 2020 Project Manager from 2013 until 2021. In this role he first initiated the concept work for the Mars 2020 mission and then headed the development and implementation team. Perseverance is the fifth Mars rover on which he has worked. He began as a power systems engineer for the Sojourner rover of Mars Pathfinder, led the assembly and test team for the twin Spirit and Opportunity MER missions and was then Flight System Manager for Curiosity. He has worked on other planetary missions at JPL and as program manager for Earth-observing satellites in the aerospace industry – in particular a number of systems and program management positions at Orbital Sciences Corporation on remote sensing satellites for several years before he rejoined JPL in 2001.

Mr. Wallace gained a B.S. in Systems Engineering upon graduating from the US Naval Academy in 1984 and received an M.S. in Electrical Engineering in 1991 from CalTech prior to joining JPL. In addition, he served five years as an officer in the US Navy fast attack submarine fleet.

Dr. Welch started his career at JPL in the Robotics Technology Section where he worked on a variety of robotic technology tasks for space and Earth applications. In 1997 he joined the Mars Pathfinder Sojourner Rover Operations Team and has been part of JPL Mars mission development and operations since then, including MER, MSL, InSight and Mars 2020.

Fig. A3.2.22 Dr. Richard V. Welch

He graduated from high school in 1981, got an associate degree in Engineering from Berkshire Community College in 1985 and then went to the University of Massachusetts, Amherst, where he got a B.S., an M.S. and Ph.D. in Mechanical Engineering. On flying to California, he fell in love with testing robot rovers in the sandbox at JPL.

Fig. A3.2.23 Dr. Kenneth Williford

Dr. Williford has been with JPL since 2012 and has served as the Deputy Project Scientist for the NASA Mars 2020 mission since 2014. He is also the Director of the JPL Astrobiogeochemistry Laboratory, which traces the flow of biologically important elements (e.g. C, H, O, N, and S) through systems. In recent years, he has focused on developing analytical techniques to seek signs of life in some of the oldest rocks on Earth. Fossils are very rare in these rocks, and those that do exist are nearly always microscopic. He therefore studies the biogeochemistry of ancient rocks, in part to understand how similar techniques can be applied to the search for evidence of life on other planets in preparation to analyze the samples of Mars that will be returned to Earth by the Mars Sample Return mission. His laboratory is part of the JPL Center for Analysis of Returned Samples. This is a cooperative joint facility for clean sample preparation and curation, microscopic characterization, and geochemical analysis of ancient terrestrial/extraterrestrial materials.

Dr. Williford got his B.S. in Natural Resources from the University of the South in Sewanee, Tennessee in 1998 and then an M.S. in Geological Sciences in 2000 and a Ph.D. in Earth and Space Sciences (Astrobiology certificate) in 2007, both from the University of Washington in Seattle.

A3.3 Perseverance Principal Investigators

Fig. A3.3.1 Dr. Abigail Allwood

Dr. Allwood is the Principal Investigator on the Mars Rover 2020 team using the Planetary Instrument for X-Ray Lithochemistry (PIXL) to search for signs of life on Mars. In fact, she is the first female and first Australian PI on a NASA Mars mission. As a field geologist whose interests focus on the early Earth, microbial sediments, evaporites and the oldest record of life on Earth, she has studied 3.43 billion year old stromatolites (microbial sedimentary structures) and is seeking to understand the formation and interpretation of traces of primitive microbial life in extremely ancient rocks. Approaches include paleo-environmental interpretation, integration of field relationships, morphology and assemblage characteristics of stromatolites, sedimentary fabrics, mineral paragenesis, Raman spectroscopy, X-ray fluorescence chemical imaging, and stable isotope geochemistry.

She earned a B.App.Sc in Geoscience with Distinction in 2001 and a B.App.Sc. (Honors 1st Class) in Geoscience in 2002, both from Queensland University of Technology in Brisbane, Australia, then a Ph.D. in Earth Science in 2006 from Macquarie University, Sydney, Australia.

Fig. A3.3.2 Dr. Luther Beegle

Dr. Beegle is the Principal Investigator for the SHERLOC instrument carried by Perseverance. He is also Deputy Manager of the Science Division at JPL, having responsibilities that include conducting NASA funded research as a PI and Co-I in planetary science focusing on detection and characterization of organic molecules for the identification of potential biosignatures.

After gaining his B.S. in Physics and Astronomy in 1990 from the University of Delaware, he got his M.S. in Physics in 1995 and Ph.D. in Astrophysics in 1997 from the University of Alabama at Birmingham.

Dr. Beegle is also the Surface Sampling System Scientist on the SASHaP of the Curiosity rover. After supporting the development of the hardware testbeds, he identified samples for ambient testing until MSL landed. He has participated in scientific operations that focused on the properties of surface materials and the acquisition and processing of samples in Gale Crater. His current primary area of interest is developing analytical instrumentation techniques for the in-situ search for organics. He also conducted astrobiological experiments to help elucidate the conditions organic molecules might face on extraterrestrial planets. Dr. Beegle is the recipient of many NASA and JPL awards, and holds a number of patents. He has been with JPL for over twenty years.

Fig. A3.3.3 Dr. James ("Jim") F. Bell III

Dr. Bell is a Professor of Astronomy at Arizona State University specializing in the study of planetary geology, geochemistry and mineralogy using data gained from telescopes and from spacecraft. He has worked on Mars Pathfinder, Near Earth Asteroid Rendezvous (NEAR), Comet Nucleus Tour (CONTOUR), Mars Odyssey, Mars Reconnaissance Orbiter, Lunar Reconnaissance Orbiter, Mars Exploration Rover and Mars Science Laboratory Curiosity. In July 2004 he was chosen by the Mars 2020 mission to be the PI of the Mastcam-Z imager for the Perseverance rover. He is the author or co-author of a number of popular books about the exploration of space. His 'Postcards from Mars' features images taken by the Mars rovers.

Dr. Bell earned a "double" B.S. majoring in Planetary Science and Aeronautics from CalTech and an M.S. and Ph.D. from the University of Hawaii in Geology and Geophysics. He currently serves on the Board of The Planetary Society and was its President 2009–2020. In 2011 he was awarded the Carl Sagan Medal for Excellence in Public Communication by the American Astronomical Society.

Fig. A3.3.4 Dr. Michael H. Hecht

Dr. Hecht is the Principal Investigator for the Mars Oxygen ISRU Experiment (MOXIE) on the Perseverance rover. He was one of the scientists who received the 2020 Breakthrough Prize in Fundamental Physics for his contribution to the Event Horizon Telescope, a network of radio antennas that managed to take the first image of a supermassive black hole in the core of a galaxy.

Hecht joined the JPL staff in 1982 where he researched microelectromechanical systems, surface and interface science, scientific instrument development, and planetary science. He co-invented the Ballistic Electron Emission Microscopy system and published several key papers on metal-semiconductor interfaces for which he received the newly renamed Lew Allen Award for Excellence in 1990. As supervisor of the In-Situ Exploration Technology Group in the Microdevices Laboratory he developed the concept for the Deep Space 2 micro-landers which piggy-backed to Mars with the Polar Lander in 1999 but were lost. He was later named the Project Manager, Co-investigator, and Project Scientist for the Mars Environmental Compatibility Assessment (MECA) instrument for the canceled Mars Surveyor 2001 mission. This instrument was later flown as the Microscopy, Electrochemistry, and Conductivity Analyzer on the Phoenix mission to Mars in 2007 with him as Lead Scientist and Principal Investigator. It was instrumental in the discovery of perchlorate in Martian soil. Based on that work Hecht published highly-cited papers on the chemistry of Martian soil and the existence of water on Mars.

Hecht obtained a B.S in Physics from Princeton University, an M.S. from MIT and a Ph.D. from Stanford University in 1982.

Fig. A3.3.5 Dr. José Antonio Rodríguez Manfredi

Dr. Manfredi is Director of the Space Instrumentation Group at the Center for Astrobiology in Madrid, Spain. His work focuses on the development of space instrumentation for environmental and geobiological characterization of solar system bodies and also the study of extreme environments on the Earth. He was Principal Investigator of the TWINS instruments on the InSight lander, Mission Manager for REMS for the Curiosity rover, and is Principal Investigator for the MEDA instrument on the Perseverance rover.

Dr. Martínez is Co-I (with Dr. Manfredi) of the MEDA instrument for the Mars 2020 rover. He defined the science requirements of the thermal infrared sensor that measures the net thermal infrared and solar radiation at the surface of Mars.

Fig. A3.3.6 Dr. Germán Martínez

As a Staff Scientist with the Lunar and Planetary Institute in Houston, Texas his Martian research focuses upon the formation of aqueous saline solutions (brine), the solar ultraviolet and atmospheric longwave radiative environment, and how the regolith interacts with the atmosphere by the exchange of energy, water and dust. He is a team member of the Mars Science Laboratory mission, focusing on the environmental conditions measured by the Rover Environmental Monitoring Station (REMS) around the clock.

He earned his B.S. in Physics in 2005, his M.S. in Atmospheric Sciences in 2007 and his Ph.D. (summa cum laude) in 2010, all at the Universidad Complutense de Madrid, Spain. He has a diverse background in data analysis, numerical modeling, instrument development and lab work.

Dr. Hamran is the Principal Investigator for the RIMFAX subsurface radar on the Perseverance rover, and also a Co-PI for the WISDOM GPR experiment that will be carried by the ESA/Roscosmos ExoMars rover.

He is a Professor of Radar Remote Sensing with the Department of Technology Systems, University of Oslo in Norway and also the head of its Center for Space Sensors and Systems (CENSSS). His academic interests include ultra-wideband radar and geophysical techniques for researching the shallow surface of planetary

Fig. A3.3.7 Dr. Svein-Erik Hamran

bodies. He is particularly interested in electromagnetic geophysical methods such as induction techniques and ground penetrating radar. He led the development of RIMFAX from the Forsvarets Forskning Institute (FFI), the Norwegian Defence Research Establishment.

He gained an M.Sc. in Physics in 1984 from the Norwegian University of Science and Technology in Trondheim and a Ph.D. in Physics in 1990 from the University of Tromsø. He is an elected member of the Norwegian Academy of Technological Sciences.

Dr. Wiens is a planetary scientist at Los Alamos National Laboratory. He is the project leader for both the ChemCam instrument aboard the Curiosity rover and the SuperCam instrument on the Perseverance rover. He has directed the US and French team operating ChemCam and interpreting the data returned from Mars. He participated a several NASA robotic ventures which include missions to the Moon, Mars, and comets, including Stardust, Mars Odyssey, Lunar Prospector, and Deep Space One. In particular, he was responsible for three instruments for the Genesis mission and he served in the capacity of Flight Payload Lead. This was the first spacecraft to return to Earth from beyond the Moon, coming home with samples of the solar wind which revealed details about the composition of the Sun.

Fig. A3.3.8 Dr. Roger Wiens

He got his Ph.D. in Radiometric Isotope Analysis at the University of Minnesota. In addition to an honorary doctorate from the University of Toulouse in France, he was Knighted by the government of France for "forging strong ties between the French and American scientific communities" and "inspiring many young, ambitious earthlings." He wrote the much-praised book 'Red Rover: Inside the Story of Robotic Space Exploration from Genesis to the Mars Rover Curiosity'.

A3.4 Ingenuity Helicopter Team

Ms. Aung is a Burmese-American engineer. Since joining JPL in 1990 she has worked on a range of projects related to space flight and the NASA Deep Space Network (DSN). She started her career in the Radio Frequency and Microwave Subsystems Section of the DSN where she developed and tested algorithms for the Block V Receiver. Then she worked on the monopulse radar systems which were used in combination with the 34 m (112 ft) antennas for the DSN. She also worked on the 240-GHz radiometer for the Microwave Limb Sounder, launched in 2004 as part of the payload of the Aura satellite of the Earth Orbiting System.

Fig. A3.4.1 The Ingenuity Helicopter Team

Fig. A3.4.2 MiMi Aung

In 2003 she was made technical group supervisor of the Guidance, Navigation, and Control Sensors Group. In this capacity she created sensor technologies for space flight missions. She became increasingly interested in autonomous space exploration and was made manager of that Section in 2010. She was appointed Deputy Manager of the Autonomous Systems Division in 2013.

Most recently, she was Lead Engineer for the Mars Helicopter Ingenuity of the Mars 2020 mission. This achieved the first extraterrestrial powered controlled flight on April 19, 2021. She is also working on the upcoming Psyche mission to investigate the asteroid of that name.

She studied engineering at the Grainger College of Engineering at the University of Illinois at Urbana-Champaign, where she received her B.E. and M.S. degrees.

Fig. A3.4.3 Dr. J. "Bob" Balaram

Dr. Balaram is a Principal Member of Technical Staff of the Mobility & Robotic Systems Section at JPL, and works on telerobotics technology development for Mars rovers, planetary balloon aerobot systems and multi-mission, high-fidelity simulators for Entry, Descent and Landing and for Surface Mobility operations. He first proposed a micro-sized helicopter about 15 years before the Mars 2020 program, but it was not funded. Many years later, Dr. Charles Elachi, who was then director of JPL, asked Dr. Balaram to submit a new proposal that this time was approved with himself as Chief Engineer. His team worked for six years to design, build, and fly the Ingenuity helicopter on Mars as part of the Mars 2020 mission.

He earned his B.Tech. in Mechanical Engineering from the Indian Institute of Technology in 1980, then attended the Rensselaer Polytechnic Institute for an M.S. in Computer & Systems Engineering in 1982 and a Ph.D. in the same in 1985.

Dr. Chahat joined JPL in 2014 as a Senior Antenna and Microwave Engineer. He has over ten years of experience in design, analysis and development of antennas ranging from UHF to THz frequencies for communications, radar, and

Fig. A3.4.4 Dr. Nacer Chahat

imaging. His expertise includes satellite communications antennas, wearable and flexible antennas for communications and biotelemetry, and antennas for remote sensing and radio astronomy. He designed the antennas for the radio system which links the Ingenuity helicopter and the Perseverance rover.

He attended the University of Rennes in France, where he earned both an M.S. in Electrical Engineering and Radio Communications (valedictorian and summa cum laude) in 2009 and a Ph.D. in Signal Processing and Telecommunication (summa cum laude) in 2012.

Dr. Håvard Fjær Grip is a Norwegian cybernetics engineer who joined the JPL Guidance and Control Analysis Group in 2013 as a robotics technologist. He is leading the Mars Helicopter Guidance, Navigation and Control team, where he designs algorithms and software to help to control and guide vehicles. As Chief Pilot for the Ingenuity helicopter, his duties were planning a flight, constructing the command sequences and analyzing the resulting flight data.

He received his M.S and Ph.D. in Engineering Cybernetics from the Norwegian University of Science and Technology in 2006 and 2010, respectively. Prior to joining JPL he performed research and development work at the Research Group in Trondheim, Norway, the Daimler automotive company in Stuttgart, Germany, and Washington State University in Pullman, Washington.

Fig. A3.4.5 Dr. Håvard F. Grip

Fig. A3.4.6 Joshua A. Ravich

Mr. Ravich is the Ingenuity Mars Helicopter Mechanical Engineering Lead, and his descriptions of its flights can be read on the JPL Helicopter Team's website.

He is a member of the Technology Infusion group at JPL where he works on the mechanical design and analysis of spacecraft systems, most recently on the Mars Helicopter. He gained his B.S. in Mechanical Engineering from the University of California, Berkeley and an M.S. in Mechanical and Aerospace Engineering from the University of Michigan.

Fig. A3.4.7 Theodore ("Teddy") Tzanetos

Mr. Tzanetos was Tactical Lead for Cruise and Surface Operations of the Mars Helicopter Ingenuity. For development he was the Assembly, Test and Launch Operations (ALTO) Lead, in which role he was the interface between the Mars 2020 team and the Ingenuity team. He was also the Test Conductor or the first flight in the test facilities at JPL. As a robotics technologist who joined JPL in 2017 to mature the level of technology readiness of such rotorcraft he has been appointed Principal Investigator for the Mars Science Helicopter (MSH) which will apply the lessons learned from Ingenuity.

Mr. Tzanetos gained a B.S. in Computer Science and Engineering from MIT in 2012 and an M.S. in Engineering in 2013. He worked at MITLL (a NYC-based startup) as Head of Technology for the Drone Racing League running a team of developers and engineers building drone-race infrastructure, embedded avionics and an online FPV simulator played by tens of thousands of users worldwide. He served as senior software engineer at Samsung Research of America.

A3.5 International Partners

While many international space agencies have supported the International Space Station for decades their support for missions to Mars has been more instrument and spacecraft oriented. Sometimes NASA contributes to their spacecraft and on occasions they contribute to ours.

For example, the Mars 2020 mission includes international contributions from Spain (INTA), France (CNES), Norway (FFI), and Italy (ASI):

- Spain's space agency, the Instituto Nacional de Técnica Aeroespacial (INTA) is providing the High Gain Antenna, the Mars Environmental Dynamics Analyzer (MEDA) instrument, and a calibration target for the SuperCam instrument.
- France's space agency, the Centre national d'études spatiales (CNES) is providing the mast unit for the SuperCam instrument.
- Norway's Forsvarets Forskning Institute (FFI) is providing the Radar Imager for Mars' Subsurface Experiment (RIMFAX) instrument.
- Italy's space agency, the Agenzia Spaziale Italiana (ASI) is providing a laser retroreflector.

Other international cooperation includes:

- NASA contributed two Electra radio communications systems to the ESA ExoMars Trace Gas Orbiter and assisted with the navigation, tracking and data return of the spacecraft.
- NASA is contributing the mass spectrometer and the main electronics for the Mars Organic Molecule Analyzer (MOMA) of the ExoMars rover to allow detection and characterization of organic molecules. The ExoMars rover is to include a drill which will allow sampling of the subsurface of Mars to a depth of almost 2 meters.
- NASA also supports navigation and tracking of the Indian Space Research Organisation (ISRO) Mars Orbiter Mission (MOM), which is currently in orbit around Mars. The two agencies have an agreement to work on future Mars missions.

- The United Arab Emirates (UAE) launched the Hope orbiter to Mars on July 19, 2020 and NASA is working to assist the mission. The spacecraft was assembled by the University of Colorado Boulder's Laboratory for Atmospheric and Space Physics (LASP) with support from Arizona State University (ASU) and the University of California, Berkeley.
- NASA is planning to help develop one of the remote sensing instruments for the Martian Moon Exploration (MMX) mission that is currently being developed for launch in 2024 by the Japan Aerospace Exploration Agency (JAXA). It will hopefully be able to bring back the first samples from the larger of Mars' two moons Phobos, and do a flyby of Deimos.
- Related to the Mars 2020 mission is the retrieval of the samples collected and cached by Perseverance. There is an ambitious campaign by NASA and ESA to collect the samples and return them to Earth, starting later in this decade.

IMAGE LINKS

Fig. A3.1.1 https://mars.nasa.gov/imgs/mars2020/missionteam.jpg
Fig. A3.1.2 https://science.nasa.gov/files/science-red/s3fs-public/styles/large/public/thumbnails/image/lori-glaze-320x480.jpg?itok=-pChMrJr
Fig. A3.1.3 https://science.nasa.gov/files/science-red/s3fs-public/styles/large/public/thumbnails/image/jeff-gramling-320x480.jpg?itok=JeT0PIpL
Fig.A3.1.4https://lh3.googleusercontent.com/hA5MGF1jbHGckEU17MXDUpow_GNwdQlKkMqKN6kK8Vfi1UuR1bNzuBrn6mR_wQgepcdThQ=s85
Fig. A3.1.5 https://www.caymancompass.com/wp-content/uploads/2016/09/Dave-Lavery-Read-Only.jpg
Fig. A3.1.6 https://www.nasa.gov/sites/default/files/meyer_hires_0.jpg
Fig. A3.1.7 https://astrobiology.nz/wp-content/uploads/2018/08/Mitch-Shulte.jpg
Fig. A3.1.8 http://fedastro.org.uk/fas/wp-content/uploads/2021/03/Tahu-George.jpg
Fig. A3.1.9 https://www.nasa.gov/sites/default/files/styles/side_image/public/thumbnails/image/jim_watzin.jpg?itok=N2Es8H7w
Fig. A3.1.10 https://science.nasa.gov/files/science-red/s3fs-public/styles/large/public/thumbnails/image/dr-thomas-zurbuchen-320x480.jpg?itok=S8NW6brn
Fig. A3.2.1 https://mars.nasa.gov/people/images/profile/2x2/baker-r-9186-profile.jpg
Fig. A3.2.2 https://simonsfoundation.imgix.net/wp-content/uploads/2019/10/07141248/ProposalFile-4.jpg?auto=format&q=90
Fig. A3.2.3 https://d2pn8kiwq2w21t.cloudfront.net/images/Dr._Robert_D._Braun_Director_for_Planetary_Sc.width-1024.jpg
Fig. A3.2.4 https://mars.nasa.gov/people/images/profile/2x2/achen-9167-profile-hi_923AB848-5B7F-43AF-8FCD6489CD52F086.jpg
Fig. A3.2.5 https://www.politico.com/news/2020/06/12/mars-2020-qa-313466
Fig. A3.2.6 https://airandspace.si.edu/sites/default/files/styles/medium_width/public/images/people/John-Grant-with-Curiosity-Rover.jpg?itok=Uy4_4M-L
Fig. A3.2.7 https://mars.nasa.gov/people/images/profile/2x2/ljandura-22764-profile-hi_496D8714-3A4B-455D-A396AEB0A118E19D.jpg
Fig. A3.2.8 https://mars.nasa.gov/people/images/profile/2x2/glkruizi-23003-profile-hi_4DE3DE2C-223F-46DC-B8DBBD3058EA5767.jpg
Fig. A3.2.9 https://mars.nasa.gov/people/images/profile/2x2/klichten-22946-profile-hi_0FD3D5E1-F560-4DFB-9A88FF8591B5A82F.jpg

Fig. A3.2.10 https://lh3.googleusercontent.com/qodEU3kUUQx1wrUk9Nnd7sQ61QXFWJlqaFZecZcOpezrHSPVioL_O7yv9ndIHZ_mFWRz0g=s89
Fig. A3.2.11 https://mars.nasa.gov/people/images/profile/2x2/smohan-22937-profile-hi_40D156F0-6351-43AB-AD6F7A501C9060A2.jpg
Fig. A3.2.12 https://www.seattleastro.org/events/1186
Fig. A3.2.13 https://ca-times.brightspotcdn.com/dims4/default/46ae7f4/2147483647/strip/true/crop/2000x1503+0+0/resize/1277x960!/quality/90/?url=https%3A%2F%2Fcalifornia-times-brightspot.s3.amazonaws.com%2F10%2F01%2F2be696dfb22550374386c607c7b5%2Fpas-1216-rovers-pg-002
Fig. A3.2.14 https://pbs.twimg.com/profile_images/537646334921424896/7oDetYt5.jpeg
Fig. A3.2.15 https://southpasadenan.com/grand-marshals-jessica-samuels-mark-swain-from-nasas-jpl-space-program/
Fig. A3.2.16 https://mars.nasa.gov/people/images/profile/2x2/nicoles-22820-profile-hi_85FA8817-C8A0-4604-8DCBF9315B3F3622
Fig. A3.2.17 https://www.kepplerspeakers.com/steltzner-a.jpg
Fig. A3.2.18 https://mars.nasa.gov/people/images/profile/2x2/vsun-22843-profile-hi_E6EB207A-7067-4F9B-8F36CE0368F627EF.jpg
Fig. A3.2.19 https://mars.nasa.gov/people/images/profile/2x2/jharris-22771-profile-hi_43D1C74C-4802-40DA-AC469A028CD50B93.jpg
Fig. A3.2.20 https://lh3.googleusercontent.com/proxy/hQOr4uke9D0SOlmzM9LNuBi2neQ_7l-jW6Sxa08FycSpCIs-b1kSYTxFFQf5kPSlP8iy4k8OgMRYhXxex9RvltzVB0X375Q
Fig. A3.2.21 https://mars.nasa.gov/people/images/profile/2x2/mwallace-22931-profile-hi_FBA818EF-545F-49E5-ACF9C1746DC6FD31.jpg
Fig. A3.2.22 https://mediad.publicbroadcasting.net/p/kuow/files/styles/x_large/public/201609/ken_williford.jpg
Fig. A3.2.23 https://mars.nasa.gov/people/images/profile/2x2/welch-22927-profile-hi_F63DA929-07C2-4D00-9C5E9439651A169C.jpg
Fig. A3.3.1 https://mars.nasa.gov/resources/25975/abigail-allwood-in-greenland/
Fig. A3.3.2 http://jimbell.sese.asu.edu/sites/default/files/styles/medium/public/2020-05/Jim%20Bell_0367_sm.jpg?itok=oDLU7ZLU
Fig. A3.3.3 https://mars.nasa.gov/layout/mars2020/images/luther-beegle-cropped.png
Fig. A3.3.4 https://mars.nasa.gov/people/images/profile/2x2/mhecht-23371-profile.jpg
Fig. A3.3.5 https://mars.nasa.gov/layout/mars2020/images/jose-manfredi-cropped.png
https://culturaccosmos.es/wp-content/uploads/ccc-jose-antonio-manfredi.jpg
Fig. A3.3.6 https://presspage-production-content.s3.amazonaws.com/uploads/1906/dr-german-martinez-lpi-usra-800w.jpg?10000
Fig. A3.3.7 https://www.mn.uio.no/its/english/people/aca/sveinerh/seh2020.jpg
Fig. A3.3.8 https://mars.nasa.gov/people/images/profile/2x2/rwiens-23234-profile-hi_708560B0-4DDC-4CD3-84BBEB0AA6D67DE9.jpg
Fig. A3.4.1 https://mars.nasa.gov/system/resources/detail_files/25145_Mars-Helicopter-Group-Photo-1600.jpg
Fig. A3.4.2 https://upload.wikimedia.org/wikipedia/commons/thumb/7/74/MiMi_Aung.jpg/330px-MiMi_Aung.jpg
Fig. A3.4.3 https://www-robotics.jpl.nasa.gov/images/people-1154.jpg
Fig. A3.4.4 https://scienceandtechnology.jpl.nasa.gov/sites/default/files/researcher/pictures/people-1249.jpg
Fig. A3.4.5 https://mars.nasa.gov/people/images/profile/2x2/hfgrip-23246-profile-hi_D0C9391E-FE7D-468D-A234C2081A02F8EE.jpg
Fig. A3.4.6 https://lh3.googleusercontent.com/pPkLpWviGjJnRfs9M_524oNiWlp96xgcMVrfEpaje5when5OIYHQr2BlcD79u90jVBLg3g=s85
Fig. A3.4.7 https://www-robotics.jpl.nasa.gov/images/people-1154.jpg

Appendix 4

The Search for Life On Mars

A book about the current Mars Perseverance mission and a future Mars Sample Return mission must address the fundamental question of life on Mars, or any other celestial body. Dr. Christopher P. McKay, an astrogeophysicist at NASA Ames Research Center, has been thinking and working on this topic his whole career. His research focuses on the evolution of the solar system and the origin of life. He is actively pursuing planning for future Mars missions including human exploration. He has been involved with polar and desert research, traveling to the Antarctic Dry Valleys, the Atacama Desert, the Arctic, and the Namib Desert to conduct research in these Mars-like environments. His research best captures the essence of the search for life, and explores the concept of life existing in a form that differs from that which we know. What follows are edited excerpts from his many articles and lectures.

> The primary goal of astrobiology is the search for evidence of life beyond the Earth. It is worth asking why is this search of interest and how can it be done? I submit that the answer to the first question is also the answer to the second question and that the answer is rooted in biochemistry. However, we've learned over the last few decades that rocks can be ejected intact from Mars by impacts and reach Earth. We presume that the process works in reverse as well, and could even extend to the outer solar system. Such rock transfers could carry microbial life – and possibly even hardy invertebrates such as tardigrades – between worlds. There may be life on Mars that shares a common ancestor with life on Earth. Hence the search for a second genesis must focus on biochemistry and genetics. Only if we find evidence of life with a different biochemistry would we possibly have found a second genius.

M. von Ehrenfried, *Perseverance and the Mars 2020 Mission*,
Springer Praxis Books, https://doi.org/10.1007/978-3-030-92118-7

On Earth, we have only one example of biochemistry. There is a common core of biomolecules that compose the hardware of life, namely proteins composed of 20 amino acids, information-bearing molecules made up from five nucleotide bases, polysaccharides constructed from a few simple sugars, and a few distinct lipids. In addition, the genetic information within organisms (the software) is common in all life. Phylogenetic trees based on that commonality indicate that all life on Earth can be traced to a common ancestor. Despite the biochemical and genetic unity between Earth organisms, there are peculiar forms (such as viruses) which have frustrated all attempts to provide a simple definition of life.

There is only one example of life on Earth and thus generalizations are not possible, yet at the same time life is too broad a phenomenon to allow for a complete characterization. To increase our difficulty, we have no consensus theory for how life was started on Earth, or whether it was carried here from elsewhere. Nor have we been able to reproduce the act in the laboratory. The desired approach to this problem of understanding the nature of life is to obtain more data. This means finding another example of life and understanding its (different) biochemistry. That is the goal of astrobiology.

Fig. A4.1 Dr. Christopher P. McKay. Photo courtesy of NASA

Second genesis

Although it is possible in principle that there is a second genesis of life hidden in some remote or extreme environment on Earth, so far every life form we've discovered on this planet maps simply on the same tree of life. It includes the life forms found deep underground, in deep sea vents, and in the most arsenic-laden, salt-rich, or acidic waters studied. So far 'life as we know it' is all that we have found on this planet. Not so long ago, it was assumed that any life that we might find on another planet would be alien and unrelated to Earth life for the simple reason that it was on another planet.

Astrobiology

It is not unusual for discussions of alien life to dwell on silicon life, or something equally beyond the domain of carbon and water which forms Earth life. However, even within the boundaries of carbon and water it is possible for life to be very different from Earth life. The simplest example that there is more than one way to make life from carbon and water is to consider the mirror image of Earth life. Many of the core biomolecules used in Earth life, and especially the amino acids possess chiral asymmetry. These molecules exist in two forms which are mirror images of each other, often referred to as the right-handed and left-handed versions. Mirror life would be alive, and its fabrication poses a well-defined challenge for synthetic biology. It should be possible to construct a simple microorganism from mirror biochemistry. This would represent a true second genesis, albeit an artificial one.

Recent research has shown that there are more possible variations in biochemistry than just the mirror image. Lab work on nucleic acids has shown that, in addition to the familiar DNA and RNA, there is what we might call 'XNA' using nucleic acid polymers composed entirely of unnatural building blocks. These have been incorporated into a simple version of the transcription machinery, which proves they are indeed viable. The actual amino acids may be different as well. Studies of the 20 or so amino acids used in proteins has led to the suggestion that approximately ten of them, the simple ones, were inherited by early life from prebiotic synthesis and that the other ten were invented by biology.

Life that originates in water is likely to use the same set of simple amino acids. This is both because they are produced prebiotically and because the number of variations is small. The simplest amino acid, glycine, contains two carbons, and it does not have left- and right-handed versions. Any life form using amino acids will use the same glycine molecule. It has been found in meteorites and comets, and it is likely to be widespread in the galaxy. There are not many choices for three carbon amino acids, hence serine and alanine are likely to be used as well. They are also found in abiotic sources. On the other hand however, the large amino acids like histidine with six carbons are not produced in prebiotic syntheses. These were chosen for biochemistry by evolution, and the possible number of variations of six carbon amino acids is huge, well over 1,000 types. Life using amino acids may well require to evolve structures with some of the functionality of histidine, but the chances of independently selecting the same molecule are small. When all ten complex amino acids are considered, the probability of a second genesis that uses amino acids in water for protein construction being based on the same 20 amino acids with the same left-handed preference is astronomically small.

I have focused on amino acids because they are a nice illustration of how carbon-based life which lives in liquid water can be distinct and different from life as we know it on Earth. The differences may be as simple as I indicated or may surprise us entirely. An analogy I use in explaining this is to think of five science books in a library. Consider the situation that one of the five books is in English, one in Spanish, one in Russian, one in Arabic, and one Chinese. All five books give directions regarding how to make a book from paper and cloth. The Spanish book shares the same alphabet with the book in English yet it is clearly of a different origin. The Russian book shares some of the same letters. The Arabic book uses an alphabet but it is one that has no symbols in common with the alphabet used for English. The Chinese book uses a system which has no simple relationship to the small number of alphabet characters of the other books. The Chinese book contains over 5,000 unique characters which represent words, ideas and sounds. Comparing these five books, they are the same at the bottom level (they are made of paper and ink) and they are the same at the highest level (they are all books describing how to make a book). Where they differ is at the level of how the ink is arranged on the paper to convey the necessary information. This is analogous with our idea of a second genesis of life occurring on another water world. Life would be the same at the bottom level (made of carbon and living in water) and would be the same at the highest ecological level (using sunlight and carbon dioxide and with big fish eating the little fish). The difference that we are seeking is the difference in the mid-levels of how biomolecules and the associated information are assembled.

This approach, searching for alien biochemistry, is the answer to the two questions I posed at the start. The search is of interest because if it is successful, then for the first time we would have two examples of how carbon chemistry can produce living systems. Our scientific understanding of how carbon-based biochemistry works is likely to be greatly advanced by having two different examples to compare. Our search shouldn't be for fossils, nor even for alien life forms as such, it should be for the biomolecules of alien life and in particular for the amino acids.

Mars, Europa and Enceladus

The worlds we can search now are those of our solar system. Three of these worlds show compelling evidence of liquid water, either in the past or at the present time. These water worlds are Mars, Europa and Enceladus. Mars is a special target because it was the first world on which evidence of liquid water was found. This evidence was the images from Mariner 9 that showed dry river channels. Subsequent data has confirmed the general picture that early in its history Mars had stable flowing liquid water on its surface. There is evidence to suggest that even now there may be limited flow of liquid water or brine. Mars shows it once had many of the diverse environments we know on Earth, including volcanoes, permafrost, ice-caps and sand dunes. For all of these reasons, Mars became the focus of the first, and to date only search for life on another world. It began with the Viking missions.

Although convincing evidence of past liquid water on the surface of Mars has been easy to come by, any indication of organics has been elusive. The Viking landers each carried a gas-chromatograph mass spectrometer that analyzed the gases released when the Martian soil had been heated to 500°C. Nothing Martian was reported. The only detection was a handful of organics at parts per billion levels which suggested they were terrestrial contamination. The Viking biology experiments detected a high reactivity in the soil consistent with a chemical oxidant. The lack of organics was unexpected, obviously, or a multimillion dollar instrument to characterize them would not have been sent. The puzzle was solved in 2008 when the Phoenix mission to Mars revealed that the chlorine in the soil on Mars was virtually all in the form of perchlorate. It was quickly realized this explained the Viking results because when perchlorate is heated to that temperature it will decompose, releasing reactive chlorine and oxygen. These destroyed any organics present and produced traces of chlorinated hydrocarbons which were erroneously attributed to contamination. Laboratory experiments have also shown that the cosmic radiation of perchlorate created reactive oxychlorides that can explain the Viking biology data. The Curiosity rover on Mars now is also defeated in its search for organics by the presence of perchlorate in the soil. (The Curiosity payload was built before the discovery from Phoenix.) Further progress in the search for alien biochemistry on Mars must await a future mission that carries an organic analyzer that doesn't rely on high temperature heating for sample processing.

Beyond Mars, Europa, one of the four large moons of Jupiter, was discovered to have liquid water by the Voyager spacecraft and this was conclusively established by the Galileo mission. However, the intense radiation environment at Europa due to Jupiter's magnetic field has stymied further missions and there has been no follow-up on the discovery of an ocean there. Like Mars, the next step would be to find some organics.

The difficulties of perchlorate on Mars and radiation on Europa are in stark contrast with the situation of Enceladus, a small, 504 km diameter moon of Saturn. Enceladus was revealed by the Cassini mission to have a plume of water emanating from its south pole. By flying through the plume, Cassini was able to show the presence of organics and salts, supporting the opinion that the plume originates from a liquid water source. Observations of the gravitational field of Enceladus also indicate an extensive liquid water ocean beneath the ice. One mission therefore discovered a liquid water environment, discovered and characterized organics from that environment and showed that the liquid water environment was habitable for Earth-like life. And samples are jetting into space, ready for the taking! The biggest challenge is bringing this material back to Earth in a manner that is consistent with the international rules on planetary protection.

Titan

The water worlds of Mars, Europa and Enceladus are of particular interest because they may contain life that is similar to life on Earth. Hopefully, if that life represents a second genesis, it would have a biochemistry that has distinct differences with Earth biochemistry. The searches on these worlds is well formed and well defined. But this similarity of environment raises the possibility that life on all these water worlds has a common origin and was spread by rocks kicked out by impacts. Where to search then?

Titan is the only solar system body other than Earth that has a stable liquid on its surface, namely a mixture of liquid methane and ethane. But it is cold

(95 K) and non-polar and therefore a very poor solvent compared with water on Earth.

Titan is also the only moon in the solar system to have a significant atmosphere, 1.5 atmospheric pressure composed of 95% molecular nitrogen with 5% methane. Solar-driven photochemistry creates an array of organic compounds, including solid organic haze particles. There is adequate chemical energy to support life in this mixture of organics. Could there be a biochemistry based on carbon but living in liquid methane? This prospect lies beyond our understanding of the possibilities for carbon biochemistry but if life were to be found on Titan it would certainly be a second genesis.

What if we succeed? The case of Mars

If we do discover a second genesis of life on Mars, does it have any implications beyond the scientific realm? Currently the international rules for planetary protection focus upon protecting future scientific investigations, not on protecting extraterrestrial organisms or their ecosystems. If we find a second genesis of life on Mars, even if the representatives of that life were only microscopic, that would raise new and profound issues in environmental ethics and would (or at least should) give us pause to think about how we should act with regard to that life.

For a 59 minute video by Dr. McKay speaking in 2013 at the opening of the UK Centre for Astrobiology at the University of Edinburgh about "Life beyond the Earth" go to: https://www.youtube.com/watch?v=VHJRUYk3cHE

For a 24 minute video on Dr. McKay speaking in 2015 to an audience at Oxford University about "The search for a second genius of life: the next 20 years" go to: https://www.youtube.com/watch?v=jHnO6tvP5PE

IMAGE LINKS

Fig. A4.1 https://web.archive.org/web/20150718171007/http://www.nasa.gov/centers/ames/research/2006/mckay.html

Appendix 5

Quotes

- *"Far better it is to dare mighty things, to win glorious triumphs, even though checkered by failure ... than to take rank with those poor spirits who neither enjoy nor suffer much, because they live in the gray twilight that knows neither victory nor defeat."*

  Theodore Roosevelt

- *"The longer you look back, the farther you can look forward."*

  Winston Churchill

- *"We're seeking signs of life, and that motivates a different suite of instruments. On the robotic arm, we have an instrument called PIXL, which measures the elemental distribution in a postage-stamp-sized area of rock. In that same area, we can take visual imagery with an instrument called WATSON. And we can measure the distribution of organic matter with an instrument called SHERLOC. These things together provide the most compelling way to find evidence of the kind of simple life that might have existed on Mars."*

  Ken Farley, Project Scientist for Perseverance at NASA's JPL.

- *"Mastcam-Z will be the main eyes of NASA's next Mars rover."*

  Jim Bell, Principal Investigator

- *"MEDA will help prepare for human exploration by providing a daily weather report and information on the radiation and wind patterns on Mars."*

  José A. Rodriguez Manfredi, Principal Investigator

- *"When we send humans to Mars, we will want them to return safely, and to do that they need a rocket to lift off the planet. Liquid oxygen propellant is something we could make there and not have to bring with us. One idea would be to bring an empty oxygen tank and fill it up on Mars."*

  Michael Hecht, MOXIE Principal Investigator

- *"If you are looking for signs of ancient life, you want to look at a small scale and get detailed information about chemical elements present."*

M. von Ehrenfried, *Perseverance and the Mars 2020 Mission*,
Springer Praxis Books, https://doi.org/10.1007/978-3-030-92118-7

"No evidence of life on Mars has ever been found. Each rover mission has inched closer to that goal, however."

"Life is not a fussy, reluctant and unlikely thing. Give life half an opportunity, and it'll run with it."

Abigail Allwood, PIXL Principal Investigator

- *"Key, driving questions are whether Mars is, or was ever inhabited, and if not, why not? The SHERLOC investigation will advance understanding of Martian geologic history and identify its past biologic potential."*

  Luther Beegle, Principal Investigator

- *"SuperCam's laser is uniquely capable of remotely clearing away surface dust, giving all of its instruments a clear view of the targets."*

  Roger Wiens, Principal Investigator

- *"No one knows what lies beneath the surface of Mars. Now, we'll finally be able to see what's there."*

  Svein-Erik Hamran, RIMFAX Principal Investigator

- *"The hardware performed as commanded but the rock did not cooperate this time. It reminds me yet again of the nature of exploration. A specific result is never guaranteed no matter how much you prepare."*

  Louise Jandura, Sampling & Caching Chief Engineer

- *"I have seen my fair share of spacecraft being lifted onto rockets but this one is special because there are so many people who contributed to this moment. To each one of them I want to say: We got here together, and we'll make it to Mars the same way."*

  John B. McNamee, Perseverance Project Manager for Mars 2020

- *"Curiosity. InSight. Spirit. Opportunity. If you think about it, all of these names of past Mars rovers are qualities we possess as humans. We are always curious, and seek opportunity. We have the spirit and insight to explore the Moon, Mars, and beyond. But, if rovers are to be the qualities of us as a race, we missed the most important thing. Perseverance. We as humans evolved as creatures who could learn to adapt to any situation, no matter how harsh. We are a species of explorers, and we will meet many setbacks on the way to Mars. However, we can persevere. We, not as a nation but as humans, will not give up. The human race will always persevere into the future."*

  Thirteen year old Alexander Mather, whose essay suggested the named that was chosen for the 'Perseverance' Mars 2020 rover

- *"As the Mars Sample Return campaign moves forward and into the next decade, and as more spacefaring nations and commercial partners emerge, our understanding of the Martian environment will become wider and deeper. We're on the cusp of profound advances in deep space exploration that will initiate the shift from robotic to human exploration of Mars."*

  Michael Meyer, lead scientist for NASA's Mars Exploration Program

- *"While we continue to maximize the science return from our two Mars orbiters (Mars Express and ExoMars Trace Gas Orbiter), we are also gearing up for a safe landing and roving across the planet's surface. To secure our future in Mars exploration, looking towards human exploration of the Red Planet, we are already planning the next logical steps; a robotic sample return mission as the first round-trip to the surface of Mars. NASA's 2020 rover mission will soon be in place as the first step of this challenging mission. Now we want to finish it.*

David Parker, ESA's Director of Human and Robotic Exploration

- *"MSR will foster significant engineering advances for humanity and advance technologies needed to successfully realize the first round-trip mission to another planet. The scientific advances offered by pristine Martian samples through MSR are unprecedented, and this mission will contribute to NASA's eventual goal of sending humans to Mars."*

Jeff Gramling, the director of the MSR program at NASA Headquarters

- *"I used to live in an orphanage. It was dark and cold and lonely. At night, I looked up at the sparkly sky and felt better. I dreamed I could fly there. In America, I can make all my dreams come true. Thank you for the 'Spirit' and the 'Opportunity'."*

Sofi Collis, a nine year old third-grade Russian-American student from Arizona who named the rovers by winning a student essay competition

- *"The ingenuity and brilliance of people working hard to overcome the challenges of interplanetary travel are what allow us all to experience the wonders of space exploration. Ingenuity is what allows people to accomplish amazing things."*

High school student Vaneeza Rupani of Northport, Alabama submitted the name 'Ingenuity' for the Mars 2020 rover, however it was selected for the helicopter to reflect how much creative thinking the team employed to get that part of the mission underway

- *"We had not originally planned to do this operational demo with the helicopter, but two things have happened that have enabled us to do it. The first thing is that originally, we thought that we'd be driving away from the location that we landed at, but the Perseverance science team is actually really interested in getting initial samples from this region that we're in right now. Another thing that happened is that the helicopter is operating in a fantastic way. The communications link is over performing, and even if we move farther away, we believe that the rover and the helicopter will still have strong communications, and we'll be able to continue the operational demo."*

Jennifer Trosper, Perseverance Project Manager

Appendix 6
Timeline

1994	Start of the Mars Exploration Program
2001	MEPAG Mars Goals, Objectives, Investigations and Priorities
7/1/13	Announcement of Opportunity for Mars 2020 investigations
7/13/13	Science Definition Team submits report
5/14–16/14	1st Landing Selection Workshop
7/31/14	NASA announces the selection of the Mars 2020 Rover science instruments
8/4–6/15	2nd Landing Site Workshop
2/8–10/17	3rd Landing Site Workshop
10/16–18/19	4th Landing Site Workshop
7/30/20	Launch of Mars 2020 to Mars
2/18/21	Perseverance arrived in Mars in Jezero Crater
2/19/21	MEDA powers up, verifies systems and sends report
3/4/21	Perseverance's drive functions tested
3/21/21	Ingenuity's cover released
4/3/21	Deployment of Ingenuity
4/3–4/21	MEDA sends full weather report

M. von Ehrenfried, *Perseverance and the Mars 2020 Mission*,
Springer Praxis Books, https://doi.org/10.1007/978-3-030-92118-7

4/9/21	Ingenuity flight test propeller software
4/19/21	First flight test of Ingenuity
4/20/21	MOXI generated 5.37 grams of O_2 from CO_2
4/22/21	2nd flight of Ingenuity. Flies 5 m high and uses color camera
4/25/21	3rd flight of Ingenuity
4/30/21	4th flight of Ingenuity. Flies 133 m (436 ft); 266 m (873 ft) roundtrip back to Wright Brothers Field
5/7/21	5th flight of Ingenuity. Lands at Airfield B
5/22/21	6th flight of Ingenuity (problem with camera)
6/1/21	Perseverance begins its first science campaign on sol 100
6/8/21	7th flight of Ingenuity
6/21/21	8th flight of Ingenuity (The "watchdog issue," a recurring issue which occasionally prevented Ingenuity from taking flight, was fixed.)
7/5/21	9th flight of Ingenuity. This flight was the first to explore areas only an aerial vehicle can, by taking a shortcut over the Séítah unit whose sandy ripples would posed a serious obstacle for the Perseverance rover
7/22–25/21	The 9th International Conference on Mars
7/24/21	10th flight of Ingenuity
7/26/21	MEPAG meeting after the conference
8/4/21	11th flight of Ingenuity
8/6/21	Perseverance takes its first sample taken from the Crater Floor Fractured Rough geological unit
8/11/21	JPL determines the sample was too crumbly to fill the tube
8/16/21	12th flight of Ingenuity to scout out Séítah
9/1/21	Perseverance drilled the hole the named Montdenier on the rock named Rochette
9/4/21	13th flight of Ingenuity; observed the geological target named Faillefeu

9/5/21	Completion of Ingenuity's Demonstration Phase; mission extended indefinitely
9/7/21	Perseverance drilled the hole named Montagnac on Rochette
9/16/21	Preflight High Spin Test of Ingenuity
9/18/21	14th flight of Ingenuity canceled owing to lower atmospheric density and a server problem
10/2–16/21	Mars-Earth conjunction-no commands sent to Perseverance or Ingenuity
10/24/21	14th Flight of Ingenuity
11/5/21	Perseverance abraded "Brac" rock
11/6/21	15th Flight of Ingenuity with 2700 RPM rotor speed
11/15/21	Perseverance took 3rd sample
11/21/21	16th Flight of Ingenuity over "Raised Ridges"
12/5/21	17th Flight of Ingenuity of 187 m (614 ft) to the Northeast
Late Dec	18th Flight of Ingenuity to the northern edge of South Séítah

References

NASA/JPL Reports

Mars Ascent Vehicle needs a Sustained Development Effort, Regardless of Sample Return Mission Timelines. John Whitehead, May 2021.

Mars Science Goals, Objectives, Investigations, and Priorities: 2020 Version. Mars Exploration Program Analysis Group (MEPAG) Goals Committee: Don Banfield, Chair. Posted March, 2020.

Creating the Future of Planetary Science (with JPL Planetary Science Directorate leadership assignments). NASA/JPL Internal Memo: Dr. Robert Braun. JPL Press Release posted: Monday, March 30, 2020.

Mars, the Nearest Habitable World :A Comprehensive Program for Future Mars Exploration. Report by the NASA Mars Architecture Strategy Working Group (MASWG), November 2020.

Mars 2020 Mission Overview. Kenneth A. Farley, JPL et al. School of Earth and Space Exploration. Arizona State University, December 3, 2020.

Tours of High-Containment and Pristine Facilities in Support of Mars Sample Return (MSR) Sample Receiving Facility (SRF) Definition Studies. NASA Tiger Team RAMA, Richard L. Mattingly, Alvin L. Smith II, Michael J. Calaway and Andrea D. Harrington. NASA Johnson Space Center, October 2020.

Mars, the Nearest Habitable World: A Comprehensive Program for Future Mars. Exploration. Report by the NASA Mars Architecture Strategy Working Group (MASWG), November 2020.

Exploring Mars with Returned Samples. Monica M. Grady, Space Science Reviews volume 216, Article number: 51 2020.

Sample Retrieval Lander Concept for a Potential Mars Sample Return Campaign. B. K. Muirhead, C. D. Edwards, Jr., A. E. Eremenko, A. K. Nicholas, A. H. Farrington, A. L. Jackman, S. Vijendran, L. Duvet, F. Beyer, S. Aziz. JPL and MSFC. European Space Agency-ESTEC, The Netherlands. Ninth International Conference on Mars 2019.

Why Mars Sample Return is a Mission Campaign of Compelling Importance to Planetary Science and Exploration. A White Paper for the Planetary Decadal Survey. Primary authors: H. Y. McSween, Kevin McKeegan, D. W.Beaty, B. L. Carrier, 2019.

M. von Ehrenfried, *Perseverance and the Mars 2020 Mission*,
Springer Praxis Books, https://doi.org/10.1007/978-3-030-92118-7

A New Era and a New Tradespace: Evaluating Earth Entry Vehicles Concepts for a Potential 2026 Mars Sample Return. Scott Perino - Jet Propulsion Laboratory, Jeremy Vander Kam - Ames Research Center, Jim Corliss - Langley Research Center. International Planetary Probe Workshop; June 11th – 15th, 2018.

The Science Process for Selecting the Landing Site for the 2020 Mars Rover. Grant, J.A., Golombek, M.P., Wilson, S.A., Farley, K.A., Williford, K.H., Chen, A., 2018.

A Look Back: The Drilling Campaign of the Curiosity Rover during the Mars Science Laboratory's Prime Mission. William Abbey, Robert Anderson, Luther W. Beegle, Noah Warner. Article September, 2018.

Searching for Life on Mars Before It Is Too Late. Alberto G. Fairén, Victor Parro, Dirk Schulze-Makuch, and Lyle Whyte. Published Online: October 1, 2017. Astrobiology Vol. 17, No. 10 Forum Articles

Groundbreaking Sample Return from Mars: The Next Giant Leap in Understanding the Red Planet. A White Paper for the NRC Planetary Science Decadal Survey, Reflecting the Viewpoints of the NASA Analysis Group CAPTEM (Curation and Analysis Planning Team for Extraterrestrial Materials). Primary Author: Allan H. Treiman, et al. Lunar and Planetary Institute, November 30, 2017.

Search for a second genesis of life on other worlds in the solar system. Christopher P. McKay, October 24, 2016.

The search for life on other worlds: Second genesis. Christopher P. McKay (NASA Ames, CA, USA), December 2014 © Biochemical Society.

SHERLOC: Scanning Habitable Environments With Raman & Luminescence for Organics & Chemicals, an Investigation for 2020. L.W. Beegle, et al. 45th Lunar and Planetary Science Conference, 2014.

Report of the Mars 2020 Science Definition Team. J.F. Mustard, chair; M. Adler, A. Allwood, D.S. Bass, D.W. Beaty, J.F. Bell III, W.B. Brinckerhoff, M. Carr, D.J. Des Marais, B. Drake, K.S. Edgett, J. Eigenbrode, L.T. Elkins-Tanton, J.A. Grant, S. M. Milkovich, D. Ming, C. Moore, S. Murchie, T.C. Onstott, S.W. Ruff, M.A. Sephton, A. Steele, A. Treiman, July 1, 2013.

A New Analysis of Mars "Special Regions". Findings of the Second MEPAG Special Regions Science Analysis Group (SR-SAG2), NNH13ZDA018O. Release Date September 24, 2013.

Announcement of Opportunity Mars 2020 Investigations. Status Report From: NASA Headquarters Science Mission Directorate. Dr. Mitchell D. Schulte. September 25, 2013, Revision Date: December 4, 2013.

Report of the Mars 2020 Science Definition Team. J.F. Mustard, chair; M. Adler, A. Allwood, D.S. Bass, D.W. Beaty, J.F. Bell III, W.B. Brinckerhoff, M. Carr, D.J. Des Marais, B. Drake, K.S. Edgett, J. Eigenbrode, L.T. Elkins-Tanton, J.A. Grant, S. M. Milkovich, D. Ming, C. Moore, S. Murchie, T.C. Onstott, S.W. Ruff, M.A. Sephton, A. Steele, A. Treiman, posted July 1, 2013 by the Mars Exploration Program Analysis Group (MEPAG).

Mars Science Laboratory Mission and Science Investigation. John P. Grotzinger, Joy Crisp, Ashwin R. Vasavada, Robert C. Anderson, Charles J. Baker, Robert Barry, David F. Blake, Pamela Conrad, Kenneth S. Edgett, Bobak Ferdowski, Ralf Gellert, John B. Gilbert, Matt Golombek, Javier Gómez-Elvira, Donald M. Hassler, Louise Jandura, Maxim Litvak, Paul Mahaffy, Justin Maki, Michael Meyer, Michael C. Malin, Igor Mitrofanov, John J. Simmonds, David Vaniman, Richard V. Welch, Roger C. Wiens. Published online: July 25, 2012.

Report of the 2018 Joint Mars Rover Mission Joint Science Working Group (JSWG). Final Version March 26, 2012.

Mars Sample Return Earth Entry Vehicle: Continuing Efforts. M. M. Munk and L. Glaab, NASA Langley Research Center, 2012.
Planning for Mars Returned Sample Science: Final report of the MSR End-to-End International Science Analysis Group (E2E-iSAG), Nov. 22, 2011.
Planning Considerations for a Mars Sample Receiving Facility: Summary and Interpretation of Three Design Studies. David W. Beaty, Carlton C. Allen, Deborah S. Bass, Karen L. Buxbaum, James K. Campbell, David J. Lindstrom, Sylvia L. Miller, and Dimitri A. Papanastassiou. 2009.
Summary of the Mars Science Goals, Objectives, Investigations, and Priorities. Mars Exploration Program Analysis Group (MEPAG) Goals Committee. Jeffrey R. Johnson, United States Geological Survey 928-556-7157, September 15, 2008.
Overview of the Mars Sample Return Earth Entry Vehicle. Robert Dillman and James Corliss. NASA Langley Research Center, 2004.
Mars Exploration Program Strategy: 1995–2020. Donna L. Shirley, Manager, Mars Exploration Program and Dr. Daniel J, McCleese, Mars Exploration Program Scientist. Jet Propulsion Laboratory, 1996.
Exploring Mars with Returned Samples. Monica M. Grady, Space Science Reviews volume 216, Article number: 51 2020.

Useful Internet Links

See the NASA Mars Exploration Program Analysis Group MEPAG site for more references.

https://mepag.jpl.nasa.gov/reports.cfm?expand=science.
http://mepag.jpl.nasa.gov/reports/MEP/Mars_2020_SDT_Report_Final.pdf.
https://mepag.jpl.nasa.gov/reports.cfm.
http://mepag.jpl.nasa.gov/reports/.
http://mepag.jpl.nasa.gov/reports/MEP/Mars_2020_SDT_Report_Final.pdf.

Lunar and Planetary Institute

http://www.lpi.usra.edu/captem/publications.shtml.

Glossary and Terminology

ACA	Adaptive Caching Assembly
ADC	Analog-to-Digital Converter
AMMOS	Advanced Multi-Mission Operations System
AO	Announcement of Opportunity
ARC	Ames Research Center
ASI	Agenzia Spaziale Italiana (Italian Space Agency)
ATR	Automatic Transfer Vehicle
BIB	Battery Interface Board
BSL-4	Biosafety Level (4 is the highest)
CAC	CO_2 Acquisition and Compression (on MOXIE)
CCD	Charge-Coupled Device
CENSSS	Center for Space Sensors and Systems (Oslo)
CETEX	Committee on Contamination by Extraterrestrial Exploration
CFFR	Crater Floor Fractured Rough (1st drill site)
CHNOPS	A mnemonic acronym for the six main chemical elements that make up living things; Carbon, Hydrogen, Nitrogen, Oxygen, Phosphorous, Sulfur
CNES	Centre National d'Etudes Spatiales (French Space Agency)
COSPAR	Committee on Space Research
COTS	Commercial Off-the-shelf
CPU	Central Processor Unit
CSIC	Spanish National Research Council
Delta-DOR	Delta-Differential One-Way Ranging (tracking and navigation)
dof	degrees-of-freedom

M. von Ehrenfried, *Perseverance and the Mars 2020 Mission*,
Springer Praxis Books, https://doi.org/10.1007/978-3-030-92118-7

DRAM	Dynamic Random-Access Memory
DSN	Deep Space Network
DUV	Deep UV (on SHERLOC)
DWI	Double-Walled Isolator
ECM	Electronics Core Module
EDL	Entry, Descent and Landing
EDM	Engineering Design Model
EEV	Earth Entry Vehicle
EEPROM	Electrically Erasable Programmable Read-Only Memory
ERO	Earth Return Orbiter
ESA	European Space Agency
ESCF	European Sample Curation Facility
ESDMD	Explorations Systems Development Mission Directorate
EURO-CARES	European Curation of Astromaterials Returned from Exploration of Space
FC	Flight Controller
FDCP	Fully Dense Carbon-Phenolic
FFB	FPGA/Flight-Controller Board
FFI	Forsvarets Forskning Institute (Norway)
FFRDC	Federally Funded Research and Development Center
FIRST	For Inspiration and Recognition of Science and Technology
FPGA	Field-Programmable Gate Array
FPV	First Person View Flight
GDRT	Gaseous Dust Removal Tool
GIF	Graphics Interchange Format
GPIO	General Purpose Input/Output
GPR	Ground Penetrating |Radar
GPS	Global Positioning System
HEO	Human Exploration and Operations
HEOMD	Human Exploration and Operations Mission Directorate
HEPA	High Efficiency Particulate Air (filter)
HiRISE	High Resolution Imaging Science Experiment (on MRO)
HS	Humidity Sensor (on the MEDA)
IC	Integrated Circuit
ICSU	International Council for Scientific Unions
IMEWG	International Mars Exploration Working Group
iMOST	International MSR Objectives & Samples Team
IMU	Inertial Measurement Unit
INTA	Instituto Nacional de Técnica Aeroespacial
IRAP	Institut de Recherche en Astrophysique et Planétologie
ISRO	Indian Space Research Organisation
JAXA	Japan Aerospace Exploration Agency
JPL	Jet Propulsion Laboratory

LaRC	Langley Research Center
LASP	Laboratory for Atmospheric and Space Physics
LED	Light-Emitting Diode (on PIXL)
LITVC	Liquid Injection Thrust Vector Control
MAHLI	Mars Hand Lens Imager (on SHERLOC)
MASWG	Mars Architecture Strategy Working Group
MAV	Mars Ascent Vehicle
MAVEN	Mars Atmosphere and Volatile EvolutioN
MCC	Micro-Context Camera (on PIXL)
MECA	Mars Environmental Compatibility Assessment
MEDA	Mars Environmental Dynamics Analyzer
MDAP	Mars Data and Analysis Program
MEDLI	Mars Science Laboratory Entry, Descent, and Landing Instrument
MEDA	Mars Environmental Dynamics Analyzer
MEMS	Microelectromechanical System
MEP	Mars Exploration Program
MEPAG	Mars Exploration Program Analysis Group
MMRTG	Multi-Mission Radioisotope Thermoelectric Generator
MER	Mars Exploration Rover (Spirit & Opportunity)
MGS	Mars Global Surveyor
MMX	Martian Moon Exploration mission (JAXA)
MOLA	Mars Orbiter Laser Altimeter
MOM	Mars Orbiter Mission (ISRO)
MOMA	Mars Organic Molecule Analyzer
MOXIE	Mars Oxygen ISRU Experiment
MRO	Mars Reconnaissance Orbiter
MSFC	Marshall Space Flight Center
MSH	Mars Science Helicopter
MSR	Mars Sample Return
MSL	Mars Science Laboratory (Curiosity)
MSPG	MSR Science Planning Group
MSSS	Malin Space Science Systems in San Diego, California,
NASA	National Aeronautics and Space Administration
NASEM	National Academies of Sciences, Engineering, and Medicine
NAV	Navigation
NRA	NASA Research Announcement
NRC	National Research Council
NSB	Navigation/Servo Controller Board
NSPIRES	NASA Solicitation and Proposal Integrated Review and Evaluation System
Pa	Pascal (The unit of pressure in the SI system)
PDS	Planetary Data System

PIXL	Planetary Instrument for X-ray Lithochemistry
PP	Planetary Protection
PSD	Planetary Science Division
R&A	Research and Analysis
RCE	Rover Computer Element
RED	Rover Equipment Deck
RIMFAX	Radar Imager for Mars' Subsurface Experiment
ROI	Region of Interest
RTE	Return-to-Earth
RSM	Remote Sensing Mast (on the MEDA)
RSSB	Returned Sample Science Board
RT	Range Trigger (EDL)
SAG	Science Analysis Group
SCF	Sample Curation Facility
SDT	Science Definition Team
SEL	Single-Event Latch-up
SFR	Sample Fetch Rover
SHA	Sample Handling Arm
SHERLOC	Scanning Habitable Environments with Raman & Luminescence for Organics & Chemicals
SMD	Science Mission Directorate
SOM	System On a Module
SOXE	Solid OXide Electrolyzer (on MOXIE)
SPI	Serial Peripheral Interface
SRF	Sample Receiving Facility
SRL	Sample Retrieval Lander
STA	Sample Transfer Arm
SRTD	Research and Technology Development
STMD	Space Technology Mission Directorate
TCB	Telecommunications Board
TCM	trajectory correction maneuver
TGO	Trace Gas Orbiter
TPS	Thermal Protection System
TRL	Technology Readiness Level
TRN	Terrain-Relative Navigation (during EDL)
UAE	United Arab Emirates
UART	Universal Asynchronous Receiver/Transmitter
USGA	US Geological Survey
WATSON	Wide Angle Topographic Sensor for Operations and eNgineering
WEB	Warm Electronics Box
WS	Wind Sensors (on the MEDA)

Terminology

Evaporites

These are layered crystalline sedimentary rocks that form from brines generated in areas where the amount of water lost by evaporation exceeds the total amount of water from rainfall and influx via rivers and streams. Their mineralogy is complex, with almost 100 varieties possible, but less than a dozen species are volumetrically important. Minerals in evaporite rocks include carbonates (especially calcite, dolomite, magnesite, and aragonite), sulfates (anhydrite and gypsum), and chlorides (particularly halite, sylvite, and carnallite), as well as various borates, silicates, nitrates, and sulfocarbonates. Evaporite deposits occur in both marine and non-marine sedimentary successions.

Fluorine

Volatiles and especially halogens (fluorine and chlorine) have been recognized as important species in the genesis and melting of planetary magmas. Data from the Chemical Camera instrument on the Mars Science Laboratory rover Curiosity now provide the first in-situ analyses of fluorine at the surface of Mars. Two principal F-bearing mineral assemblages are identified. The first is associated with high aluminum and low calcium contents, in which the F-bearing phase is an aluminosilicate. It is found in conglomerates and may indicate petrologically evolved sources. This is the first time that such a petrological environment is found on Mars. The second is represented by samples that have high calcium contents, in which the main F-bearing minerals are likely to be fluorapatites and/or fluorites. Fluorapatites are found in some sandstone and may be detrital, while fluorites are also found in the conglomerates, possibly indicating low temperature alteration processes.

Glycine

The presence of organic matter in lacustrine (lacustrine deposits are sedimentary rock formations that formed in the bottom of ancient lakes) mudstone sediments at Gale Crater was revealed by the Mars Science Laboratory rover Curiosity, which also identified smectite clay minerals (smectite and vermiculite minerals are often referred to as "swelling" or "expandable" clay minerals).

Analogue experiments on phyllosilicates formed under low temperature aqueous conditions have illustrated that these are excellent reservoirs to host organic compounds against the harsh surface conditions of Mars.

The discovery of amino acids such as glycine on meteorites and comets confirms the role of small bodies as transport and delivery vehicles of building blocks of life on Earth and possibly on other planetary bodies of our solar system. Glycine is quite interesting because it is the simplest of the 20 biogenic amino acids, from which complex organic molecules might have originated in our evolved solar system.

Paragenesis

Paragenesis is a petrological concept meaning an equilibrium sequence of mineral phases. It is employed in the study of igneous and metamorphic rock genesis and importantly in studies of the hydrothermal deposition of ore minerals and the rock alteration associated with ore mineral deposits.

Perchlorates

Perchlorates have been identified on the surface of Mars. This has prompted speculation of what their influence would be on habitability. When irradiated with a simulated Martian ultraviolet flux, perchlorates become bacteriocidal. At concentrations associated with Martian surface regolith, vegetative cells of Bacillus subtilis in Martian analogue environments lost viability within minutes. Two other components of the Martian surface, iron oxides and hydrogen peroxide, act in synergy with irradiated perchlorates to cause a 10.8-fold increase in cell death when compared to cells exposed to ultraviolet radiation after 60 seconds of exposure. These data show that the combined effects of at least three components of the Martian surface, activated by surface photochemistry, render the present-day surface more uninhabitable than previously thought and demonstrate the low probability of survival of biological contaminants released from robotic and human exploration missions.

Petrology

The study of rocks in terms of their composition, texture, and structure, their occurrence and distribution, and their origin in relation to physicochemical conditions and geological processes. It comprises three subdivisions: igneous, metamorphic, and sedimentary petrology.

Stromatolites

We do not know what Martian biosignatures might look like, but the ancient Earth might provide clues. A record of our planet's early life can be found in stromatolites; rocks originally formed by the growth of layer after layer of bacteria. If similar structures exist on Mars, scientists could combine measurements from different instruments to assess the likelihood of a biological origin.

Olivine

Olivine is a mixed crystal of Mg_2SiO_4 and Fe_2SiO_4, in which the magnesium member is usually dominant. It is the major mineral in the mantle of the Earth. Plate-tectonic processes have pushed up enormous slabs of olivine-rich rocks from the mantle to the Earth's surface, where they can be mined in open pits. Spectra of the Martian surface from the Thermal Emission Spectrometer (TES) onboard Mars Global Surveyor have been matched with laboratory spectra of olivine. It is known to crystallize first from a magma and to weather first in the presence of water into clays or iron oxides. The occurrence of olivine on the surface of Mars and its susceptibility to chemical weathering has geochemists busy investigating how long it has been there and what that means about climate history.

Ophiolite

An ophiolite is a section of Earth's oceanic crust and the underlying upper mantle that has been uplifted and exposed above sea level, often having been pushed onto continental crustal rocks. The name ophiolite means "snakestone" from "ophio" (snake) and "lithos" (stone) in Greek.

Range Trigger

The key to the new precision landing technique is choosing the right moment to pull the "trigger" that releases the spacecraft's parachute. "Range Trigger" is the name of the technique that Mars 2020 used to time the parachute's deployment. Earlier missions deployed their chutes as early as possible, after the spacecraft reached a desired velocity. Mars 2020's Range Trigger deployed the parachute based on the spacecraft's position relative to the desired target. That meant the parachute would be deployed early or later, depending upon how close it was to its desired target. If the spacecraft were going to overshoot the landing target, the parachute would be deployed earlier. If it were going to fall short of the target, the parachute would be deployed later to allow the spacecraft to fly a little nearer to its target.

Terrain-Relative Navigation

Using Terrain-Relative Navigation, the Perseverance rover can estimate its location while descending through the Martian atmosphere on its parachute. That allows the rover to determine its position relative to the ground with an accuracy of about 40 m (130 ft) or better. Until now, many of the potential landing sites on Mars have been off-limits. The risks of landing in challenging terrain were much too great. For past Mars missions, 99% of the potential landing area (the landing ellipse) had to be free of hazardous slopes and rocks to help ensure a safe landing. Using TRN, the Mars 2020 mission site selectors were able to consider more and more interesting landing sites with far less risk.

About the Author

Manfred "Dutch" von Ehrenfried started working with scientists in the Apollo days when he was the Chief of the Science Requirements and Operations Branch at NASA JSC. This Branch was responsible for the definition, coordination and documentation of science experiments assigned to Apollo and Skylab missions. This included the Apollo Lunar Science Experiment Packages (ALSEP) left on the Moon and the experiments in lunar and Earth orbit. The ALSEPs included seismic sensors, magnetometers, spectrometers, ion detectors, heat flow sensors, charged particle and cosmic ray detectors, gravity measurements and more. The lunar orbit experiments included the Scientific Instrument Module (SIM) Bay cameras and sensors and the Particles and Fields subsatellites that were released prior to leaving lunar orbit. The work also defined the astronauts' procedures for deploying the packages and conducting experiments on the Moon and in lunar orbit. He also spent one year with a contractor at NASA Goddard on the Earth Resources Technology Satellite (ERTS), later named Landsat 1. He later worked with scientists on the Hi-Altitude Earth Resources Aircraft Program. He assisted scientists with their research by planning a mission and operationally achieving their test objectives as the first sensor equipment operator and mission manager on the NASA/USAF high altitude RB-57F. These flights required wearing a full pressure suit as these missions generally flew at altitudes in the range 65,000 to 67,000 ft, with one flight actually achieving 70,000 ft.

Dutch had the very good fortune to have interviewed with the NASA Space Task Group the day before Alan Shepard was launched on MR-3. At the time, he had very little knowledge of Project Mercury and thought that since his degree was in physics he would be working in that area. As fate would have it, he was assigned to the Flight Control Operations Section under Eugene F. Kranz, who became his supervisor and mentor. Most of his work for Project Mercury was in

M. von Ehrenfried, *Perseverance and the Mars 2020 Mission*,
Springer Praxis Books, https://doi.org/10.1007/978-3-030-92118-7

Fig. AA.1 The author in late 1961 as a young STG Flight Controller. Center left: At the console to the left of Gene Kranz and George Low. Center right: Testing Neil Armstrong's suit to an equivalent altitude of 400,000 ft in the vacuum chamber at the Manned Spacecraft Center. Bottom: Wearing the A/P22S-6 full pressure suit required for the RB-57F. All photos courtesy of NASA

the areas of mission rules, countdowns, operational procedures and coordination with remote tracking station flight controllers. In training to be a flight controller he spent the MA-4 and MA-5 missions at NASA Goddard learning communications between the Mercury Control Center and the Manned Space Flight Network. After that he had the Operations and Procedures flight control console in the Mercury Control Center for John Glenn's historic MA-6 orbital flight, then went on to support the remaining missions.

After the Space Task Group relocated from Langley in Virginia to Houston in Texas, Dutch supported the Gemini missions and was Assistant Flight Director for Gemini 4 to Gemini 7, which including the first spacewalk by Ed White and the first rendezvous in space by Gemini 6 with Gemini 7. In 1966, he became a Guidance Officer on Apollo 1 and after the accident and stand-down he became the Mission Staff Engineer on Apollo 7 and backup on Apollo 8. Over this same period, he served as an Apollo Pressure Suit Test Subject. This afforded him the opportunity to test pressure suits in the vacuum chamber to over 400,000 ft. On one occasion he tested Neil Armstrong's suit. He also experienced 9 g's in the centrifuge and flew in the zero-g aircraft. He had his own Apollo A7LB Skylab suit.

Dutch also worked in the nuclear industry for seven years and wrote the book 'Nuclear Terrorism – A Primer'. He worked as a contractor with the original NASA Headquarters Space Station Task Force for ten years. He has written a number of books about his experiences, which can be seen at www.dutch-von-ehrenfried.com. For the past 25 years he has been working in the finance and insurance fields.

Index

M. von Ehrenfried, *Perseverance and the Mars 2020 Mission*,
Springer Praxis Books, https://doi.org/10.1007/978-3-030-92118-7